SOME ELEMENTARY GAUGE THEORY CONCEPTS

Born in Guangzhou, China, and brought up mostly in Hong Kong, Dr Chan Hong-Mo obtained his PhD in Birmingham, England. Before joining the Rutherford Appleton Laboratory and settling in Oxford, he was on the staff of CERN, Geneva, for nearly ten years, besides having been a member of the Institute for Advanced Study, Princeton, and a visiting professor at SUNY Stony Brook and the University of Washington, Seattle. His research interest in particle physics includes both theoretical and phenomenological topics.

Dr Tsou Sheung Tsun obtained her BSc in Hong Kong and her Doctorat ès Sciences in Geneva. She has held research fellowships at Wadham College, Oxford, and at the Mathematical Institute, Oxford, where she is now a faculty member. Trained both as a mathematician and a physicist, Dr Tsou at present works mostly in gauge theory, string theory and twistor theory.

The two authors' joint work on this volume about elementary gauge theory concepts has been a labour of love to which they have devoted a number of years.

World Scientific Lecture Notes in Physics – Vol. 47

SOME ELEMENTARY GAUGE THEORY CONCEPTS

Chan Hong-Mo
Rutherford Appleton Laboratory, UK

Tsou Sheung Tsun
Oxford University, UK

World Scientific
Singapore • New Jersey • London • Hong Kong

Published by

World Scientific Publishing Co. Pte. Ltd.

P O Box 128, Farrer Road, Singapore 9128

USA office: Suite 1B, 1060 Main Street, River Edge, NJ 07661

UK office: 73 Lynton Mead, Totteridge, London N20 8DH

SOME ELEMENTARY GAUGE THEORY CONCEPTS

ISBN 981-02-1080-9
ISBN 981-02-1081-7 (pbk)

Printed in Singapore.

Preface

Gauge theory, which underlies modern particle physics as well as the theory of gravity, and hence all of physics as we know it today, is itself based on a few fundamental concepts, the consequences of which are often as beautiful as they are deep. Unfortunately, in view of the pressure to cover aspects of the theory that are necessary for its many important applications, very little space is usually devoted in textbooks and graduate courses to the treatment of these concepts. The present little volume is an attempt to help in some small degree to redress this imbalance in the literature.

Our aim is to make these concepts and some of their immediate consequences accessible to all physicists, including graduate students, and no effort has been spared to make the whole volume understandable to the nonspecialist, given sufficient effort. The reader is warned, however, that the topics covered are elementary only in the sense of being fundamental, not in the sense of being shallow or easy. Although all will already feature at the classical field level, and most even before the introduction of an action principle, they often lead one to pose questions of some profundity which are still at the forefront of research or still unresolved. We believe, however, and we hope the reader will agree, that the subject matter is of sufficient interest to merit the effort in unravelling it. The selection of topics included may be a little personal, but for this we proffer no apology, for it is only those subjects on which we ourselves have worked do we feel confident in presenting with sufficient clarity. Obviously, this little volume is not meant to be a comprehensive work by any means. To us, it is a labour of love, a pleasure to share with others what we think we have understood, and a challenge to explain in simple terms what are at times rather difficult abstract ideas.

Our approach is physical but we shall have no hesitation in introducing mathematics when it is a help to the understanding. In the presentation, little previous knowledge is assumed of the reader apart from what is normally taught in graduate schools, but neither has there been any conscious attempt to avoid or trivialize essential difficulties and intrinsically abstruse concepts.

Since the material contains the wisdom of many, accumulated over decades, it is difficult, if not altogether impossible, to ascribe to individuals any but just some few specific items. Rather than attempting to do so, therefore, we shall quote references only chapter by chapter and only in general terms, giving not a comprehensive list, of which we are incapable, but only a list of those few particular works from which we ourselves have learned the most. We apologize to those contributors whom we should have but have not quoted; the omission is due not to wilful neglect but to ignorance on our part.

We are deeply indebted to Professor C.N. Yang for first interesting us and then in teaching us much of what we know in the subject. We have also greatly benefited from Professor G. Segal for advice, especially on the mathematics of loop space, and from Dr. P. Scharbach for a most enjoyable collaboration on monopole dynamics. Further thanks are due to Dr. R. Newman and Ms J. Faridani for a critical reading of parts the manuscript, to Mr. S. Williams for help in LaTeX, and lastly to our publisher World Scientific for their indulgence in extending our deadline time after time.

Chan Hong-Mo
Tsou Sheung Tsun

Oxford, February 1993

N.B. We use the metric $(+ - - -)$ and put $\hbar = c = 1$.

Contents

Chapter 1

Basic Concepts

1.1 Electromagnetism

The first and simplest local gauge theory known to physicists is electromagnetism, or, more precisely the theory of a charged quantized particle moving in a (classical) electromagnetic field, and in order to clarify the basic concepts of local gauge invariance, it is often rewarding to turn to this simple example for reference. In this theory, a particle is described by a complex wave function with a phase, which depends in general on the coordinates x of the particle. This phase, however, has no real physical significance, since one is allowed to redefine this phase by an arbitrary phase rotation independently at every space-time point without altering any of the physics. That such is the case is what is meant by local gauge invariance of the theory.

In view of this arbitrariness in the local definition of the phase, how can one compare the phases at two different space-time points? One can do so as follows. Introduce a set of functions $a_\mu(x)$, depending on the space-time coordinates x and labelled by an index μ running over the 4 space-time directions, say $\mu = 0, 1, 2, 3$. We then stipulate that the phase of the wave function representing a particle of charge e at a point x is parallel to the phase at a neighbouring point $x + dx^\mu$ if the local values of the phase at these two points differ by an amount $ea_\mu(x)dx^\mu$. This statement depends of course on the local definition of the phase. In order to give it an invariant meaning under local phase rotations, the functions $a_\mu(x)$ must transform accordingly. Suppose then we perform a *gauge transformation* by rotating the phase of the wave function at x by an amount $e\alpha(x)$ depending on x, namely:

$$\psi(x) \to \psi'(x) = \exp\, ie\alpha(x)\, \psi(x). \qquad (1.1.1)$$

It is clear that whereas the change in phase at x due to this transformation is $e\alpha(x)$, the phase at the neighbouring point $x + dx^\mu$ will change by an amount

1

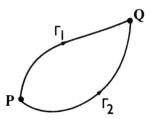

Figure 1.1: Parallel phase transport along two different paths.

$e\{\alpha(x) + \partial_\mu\alpha(x)dx^\mu\}$. Hence, in order that in the new phase convention, parallelism of phases between the two neighbouring points has still the same meaning as in the old convention, we must define the quantity $a'_\mu(x)$ after the transformation as:

$$a'_\mu(x) = a_\mu(x) + \partial_\mu\alpha(x). \tag{1.1.2}$$

The above definition of parallelism allows us to parallelly transport a wave function from any point in space-time over an infinitesimal distance in any direction. By applying the criterion repeatedly, we can then also parallelly transport a wave function over finite distances along any path. Thus, by parallel transport from a point P to a point Q a finite distance away along a path Γ, a wave function acquires a change in the local value of the phase by an amount represented by the line integral:

$$e\int_{\Gamma}^{Q}{}_{P} a_\mu(x)\,dx^\mu. \tag{1.1.3}$$

Notice, however, that the prescription above, though allowing the parallel transportation of phases from any point in space-time to any other point, will lead in general to different answers along different paths. Indeed, the difference between the phases at Q obtained by parallel transport along two distinct paths Γ_1 and Γ_2, as depicted in Figure 1.1, is just:

$$e\int_{\Gamma_2}^{Q}{}_{P} a_\mu(x)\,dx^\mu \; - \; e\int_{\Gamma_1}^{Q}{}_{P} a_\mu(x)\,dx^\mu \; = \; e\oint_{\Gamma_2-\Gamma_1} a_\mu(x)\,dx^\mu, \tag{1.1.4}$$

which is in general non-zero, depending on the functions a_μ. Indeed, using Stokes' theorem, we can rewrite the line integral on the right in (1.1.4) as a surface integral:

$$-e\iint f_{\mu\nu}(x)\,d\sigma^{\mu\nu} \tag{1.1.5}$$

over any surface Σ bounded by the closed curve $\Gamma_2 - \Gamma_1$, with:

$$f_{\mu\nu}(x) = \partial_\nu a_\mu(x) - \partial_\mu a_\nu(x). \qquad (1.1.6)$$

Hence, one sees that the phase at Q obtained by parallel transport from P would be independent of the path if the vector function $a_\mu(x)$ is curl-free everywhere. If, on the other hand, $f_{\mu\nu}$ is non-zero, then the parallel transport of phases will in general be path-dependent.

The quantity $-e f_{\mu\nu}(x) dx^\mu dx^\nu$ is the difference in phase obtained by parallel transport along respectively the paths ABC and ADC in Figure 1.2 or equivalently, it is the change in phase on returning to the point A after parallelly transporting once around the infinitesimal parallelogram $ABCD$ in the counter-clockwise direction. It thus follows that the tensor $f_{\mu\nu}$ is gauge invariant, since, being the difference of two phases both taken at the same space-time point, it is independent of any phase rotations either at that point or at neighbouring points. That this is indeed the case can of course be checked algebraically by applying (1.1.2) to (1.1.6).

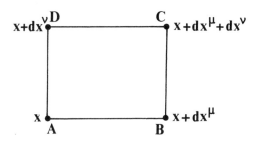

Figure 1.2: Elemental parallelogram on the $\mu\nu$-plane.

Now the change in phase under parallel transport, apart from a numerical factor which we call the charge of the particle, acts on all wave functions in the same way, and is thus not an attribute of the particle, but represents rather a physical condition of the domain of space-time under consideration. In conventional language, it is described as a field over the domain. The vector field $a_\mu(x)$ is known, of course, as the gauge potential, and $f_{\mu\nu}(x)$ is the electromagnetic field tensor.

Historically, the concept of the electromagnetic field was first introduced in classical physics where the phases of wave functions did not arise. The electromagnetic field at any point x was defined as the force that a classical charge would experience when placed at that point, while the electromagnetic

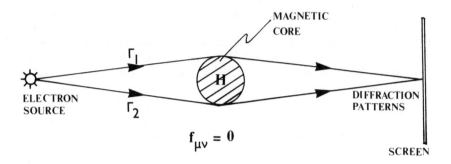

Figure 1.3: Schematic illustration for the Aharonov–Bohm experiment.

potential a_μ was introduced merely by (1.1.6) as a subsidiary quantity for convenience of mathematical manipulations. One sees now, however, that when the charged particle is quantized, these quantities take on an additional significance. It is then important to ascertain whether this newly acquired significance of the electromagnetic field is 'real' in the sense of being physically observable, and whether it is consistent with the original meaning ascribed to the electromagnetic field. That this is so was demonstrated by the famous experiment devised by Aharonov and Bohm in 1959, and first successfully performed by Chambers in 1960.

This experiment for checking the basic gauge structure of electrodynamics is one of those really fundamental experiments in physics, which is as beautifully simple in its conception as its implications are profound. The arrangement is symbolically as depicted in Figure 1.3, where electrons from a single source are allowed to pass on either side of a magnetized core and impinge on a screen behind. The magnetic field is restricted to within the core at the centre so that in the region traversed by the electrons, the field tensor $f_{\mu\nu}$ is identically zero. In classical mechanics, therefore, there will be no interaction of the electrons with the field. In the quantum theory, however, the electrons are described by a wave function with a phase, which according to the discussion in the preceding paragraphs, depends on the gauge potential in the region of space through which the electrons pass. More precisely, the phase difference obtained on the screen between the electrons arriving there along the two different paths Γ_1 and Γ_2, is, according to (1.1.4) and (1.1.5), just the total magnetic flux flowing through the core. Notice that, in contrast to the classical case, the electrons *are* affected by the field inside the core, for although the field $f_{\mu\nu}$ is everywhere zero along the paths of the electrons, the gauge potential a_μ need not be, and the phase of the electron wave function *is* affected by the value of the gauge potential.

Now, although it is not physically meaningful to compare phases at different space-time points, the theory being invariant under local phase rotations, the relative phase at the *same* point in space-time is a physical observable quantity. Indeed, the electrons arriving at the same point on the screen along the two paths Γ_1 and Γ_2 will interfere constructively or destructively depending on whether they are in phase or out of phase as calculated by the formula (1.1.5). The effect of the magnetic field H in the core will therefore show up as an interference pattern on the screen, the study of which will give a quantitative test for the validity of the formula. That the formula indeed holds has been shown conclusively by Chambers and others under a variety of experimental conditions. The positive outcome of this experiment is widely regarded as the strongest single piece of evidence supporting the basic tenets of electromagnetism as a gauge theory.

1.2   Yang–Mills Theory

Yang–Mills theory, as originally proposed in 1954, is a generalization of electromagnetism in which the complex wave function of a charged particle is replaced by a wave function with two components, say $\psi = \psi^i(x), i = 1, 2$. By a change of 'phase', we mean now a change in the orientation in *internal space* of ψ under a transformation:

$$\psi \rightarrow S\,\psi \qquad (1.2.1)$$

where S is a unitary matrix with unit determinant, so that local gauge invariance is now to be interpreted as the requirement that physics be unchanged under arbitrary $SU(2)$ transformations on ψ independently at different space-time points.

In order to specify 'parallelism of phases' at neighbouring space-time points, one introduces, by analogy with electromagnetism, a set of matrix-valued functions[1] $A_\mu(x)$, with the stipulation that the 'phase' at a point x is to be regarded as parallel to the 'phase' at a neighbouring point $x + dx^\mu$ if the local values of the 'phase' at these two points differ by the amount $gA_\mu(x)dx^\mu$ for a particle of 'charge' g. Such a 'phase' difference is obtainable from an infinitesimal transformation on the wave function: $\exp ig A_\mu(x)dx^\mu \approx 1 + ig A_\mu(x)dx^\mu$,

---

[1]To make the notation clearer, we introduce the convention that the lower case letters a_μ and $f_{\mu\nu}$ refer to abelian theory, while the corresponding upper case letters A_μ and $F_{\mu\nu}$ refer to nonabelian theory.

which will be unitary with unit determinant, as was required above, if the matrix $A_\mu(x)$ is traceless and hermitian, or in other words, if it is an element of the algebra $\mathfrak{su}(2)$.

The fact that we are dealing now with matrices which do not in general commute means that the formulae deduced above for electromagnetism will be modified. First, consider what happens to the gauge potential under a local gauge transformation, namely under a redefinition of the 'phase':

$$\psi(x) \rightarrow \psi'(x) = S(x)\psi(x) \qquad (1.2.2)$$

where S is an element of $SU(2)$ which depends in general on x. Before the transformation, we ascertained that ψ at x was parallel to:

$$\hat{\psi}(x) = \exp ig A_\mu(x)dx^\mu \, \psi(x) \qquad (1.2.3)$$

at $x + dx^\mu$. After the transformation, ψ at x is changed to ψ' in (1.2.2), while at $x + dx^\mu$, $\hat{\psi}$ is changed to:

$$\hat{\psi}'(x) = S(x + dx^\mu) \, \hat{\psi}(x); \qquad (1.2.4)$$

yet by gauge invariance, ψ' and $\hat{\psi}'$ ought still to be regarded as being parallel. However, in the new 'phase' convention denoted by prime, we can also define parallelism in terms of a new set of functions $A'_\mu(x)$ such that:

$$\hat{\psi}'(x) = \exp ig A'_\mu(x)dx^\mu \, \psi'(x). \qquad (1.2.5)$$

Hence, equating the two expressions (1.2.4) and (1.2.5) for $\hat{\psi}'$ and substituting in them (1.2.3) and (1.2.2), we have

$$\exp ig A'_\mu(x)dx^\mu \, S(x) \, \psi(x) = S(x + dx^\mu) \, \exp ig A_\mu(x)dx^\mu \, \psi(x) \qquad (1.2.6)$$

which, being valid for all ψ, holds also with ψ omitted. On expanding then to leading order in dx^μ, we have the transformation law for $A_\mu(x)$:

$$A'_\mu(x) = S(x) \, A_\mu(x) \, S^{-1}(x) - (i/g)\partial_\mu S(x) \, S^{-1}(x). \qquad (1.2.7)$$

When S is infinitesimal, say, $S(x) \approx (1 + ig\Lambda(x))$, then to leading order in Λ:

$$A'_\mu(x) = A_\mu(x) + \partial_\mu \Lambda(x) + ig[\Lambda(x), A_\mu(x)] \qquad (1.2.8)$$

which ought to be familiar.

Next, consider the field tensor, which, by analogy with the electromagnetic case in the last section, is to be defined as the difference in 'phase', suitably normalized, obtained by parallel transport along respectively the paths ABC and ADC in Figure 1.2. Let us work out explicitly what this ought to be in the case where the gauge potential $A_\mu(x)$ is a matrix. Parallelly transporting any $\psi(x)$ first from A to B yields $\exp ig A_\mu(x) dx^\mu\ \psi(x)$, then from B to C yields:

$$\psi_{ABC}(x) = \exp ig A_\nu(x + dx^\mu) dx^\nu\ \exp ig A_\mu(x) dx^\mu\ \psi(x), \qquad (1.2.9)$$

where we note that the argument of A_ν in the first factor on the right is $x+dx^\mu$, being the coordinate of B from which the second parallel transport is effected. Similarly, transporting along ADC yields:

$$\psi_{ADC}(x) = \exp ig A_\mu(x + dx^\nu) dx^\mu\ \exp ig A_\nu(x) dx^\nu\ \psi(x). \qquad (1.2.10)$$

Expanding both expressions to leading order in $dx^\mu dx^\nu$ and taking their difference, we have:

$$[ig\{\partial_\mu A_\nu(x) - \partial_\nu A_\mu(x)\} + (ig)^2\{A_\nu(x) A_\mu(x) - A_\mu(x) A_\nu(x)\}] dx^\mu dx^\nu. \quad (1.2.11)$$

Defining, as in the abelian case, $F_{\mu\nu}(x)$ to be such that $-g F_{\mu\nu}(x) dx^\mu dx^\nu$ is the difference in 'phase' obtained by parallel transport along these two paths gives for the field tensor the standard expression:

$$F_{\mu\nu}(x) = \partial_\nu A_\mu(x) - \partial_\mu A_\nu(x) + ig[A_\mu(x), A_\nu(x)]. \qquad (1.2.12)$$

By construction, $F_{\mu\nu}(x)$, like the abelian field tensor, is the difference between two 'phases' at the same space-time point, and can thus depend only on the 'phase' convention at that particular point. However, in contrast to the abelian case, 'phase' now actually means direction in internal symmetry space so that $F_{\mu\nu}(x)$ can change in direction under a change in 'phase' convention. In other words, we expect $F_{\mu\nu}(x)$ now to be *gauge covariant*, namely to transform under the transformation (1.2.2) as: [*]

$$F'_{\mu\nu}(x) = S(x)\, F_{\mu\nu}(x)\, S^{-1}(x), \qquad (1.2.13)$$

as can be readily checked also by direct computation using (1.2.7).

Further, by repeated application of (1.2.3), one can of course parallelly transport a wave function over finite distances along any path as we did in Section 1.1. However, A_μ being a matrix here, the procedure becomes a little more involved. From parallel transport over each infinitesimal segment, one gains, according to (1.2.3), an exponential factor. If A_μ were a number, like a_μ, then we can just write the product of such exponentials as an exponential of the sum of the exponents, obtaining thus:

$$\exp ig \int_\Gamma A_\mu(x)\, dx^\mu \qquad (1.2.14)$$

as the result of parallelly transporting ψ along the path Γ; this is the same as saying that there is a phase change by an amount equivalent to (1.1.3). However, now that $A_\mu(x)$ is a matrix-valued function whose values at different space-time points need not commute, the product of exponentials can no longer be written as an exponential of the sum of the exponents, since for $[A, B] \neq 0$:

$$e^A e^B \neq e^{A+B}. \qquad (1.2.15)$$

By convention, one still writes the result of parallel transport along Γ symbolically as:

$$P \exp ig \int_\Gamma A_\mu(x)\, dx^\mu, \qquad (1.2.16)$$

where P denotes *path-ordering*, meaning just a product of the exponentials for the infinitesimal segments, ordered from right to left as one moves along the path Γ. It is *not* the exponential of the line integral. This difference with electromagnetism is nontrivial and fundamental and accounts for much of the added complications in dealing with Yang–Mills theory. For example, we recall that in electromagnetic theory, we could apply Stokes' theorem to deduce that the total change in phase due to parallel transport along a closed circuit is an integral of the field tensor over *any* surface bounded by that circuit, from which result follows the immensely useful concept of electromagnetic flux. However, in the Yang–Mills theory, this no longer applies. Of course, Stokes' theorem still holds, relating the line integral of A_μ to a surface integral of its curl: $\partial_\nu A_\mu - \partial_\mu A_\nu$, but these quantities bear no direct relationship to the physically interesting phenomenon of parallel transport, which are given in terms of the quantities (1.2.12) and (1.2.16). For this reason, indeed, there does not seem to be any useful generalization of the electromagnetic concept of flux to Yang–Mills theory, which is an important question to which we shall have occasion to return later.

Such differences with electromagnetism, however, do not invalidate in any way the prescription for parallel transport which has been detailed above. In principle, therefore, one can construct gedanken experiments of the Aharonov–Bohm–Chambers type to probe the fundamental gauge nature of a Yang–Mills theory. However, it turns out that in the physical situations to which Yang–Mills theory has been applied, notably in electroweak theory and in chromodynamics, the conditions are such that no practical experiments of this type have so far been devised which can at present be performed.

For future reference, we note that under a gauge transformation, the phase factor (1.2.16) transforms as:

$$P \exp ig \int_{x_1}^{x_2} A_\mu(x)\, dx^\mu \longrightarrow S(x_2) \left\{ P \exp ig \int_{x_1}^{x_2} A_\mu(x)\, dx^\mu \right\} S^{-1}(x_1), \quad (1.2.17)$$

as can be verified by repeated application of (1.2.6). This assertion is conceptually obvious since (1.2.16) is the parallel transport from, say, x_1 to x_2 and ought therefore to be affected by the change in definition of the 'phase' at the two end-points.

1.3 Dirac Phase Factors

An interesting conceptual question to ask at this stage is: What are the actual physical quantities which describe a gauge theory?

Classical electrodynamics, as conceived by Faraday and Maxwell, can be described entirely in terms of the electromagnetic field tensor $f_{\mu\nu}(x)$. Thus, once given $f_{\mu\nu}(x)$ at a point x, we know via the Lorentz equation:

$$m\frac{d^2x_\mu}{d\tau^2} = -ef_{\mu\nu}(x)\frac{dx^\nu}{d\tau}, \tag{1.3.1}$$

exactly how a charged particle placed at x will behave. This is no longer true, however, in the quantum theory; as demonstrated by the Aharonov–Bohm experiment, the knowledge of $f_{\mu\nu}$ throughout the region traversed by the electron was still insufficient for determining the phase of the electron wave function, without which our description of the dynamical system would be incomplete. In other words, as dynamical variables, $f_{\mu\nu}(x)$ under-describe the quantum theory of a charged particle moving in an electromagnetic field.

For this reason, we have employed so far the gauge potential $a_\mu(x)$ as variables; it was found sufficient to give a full description of the physical phenomena. However, $a_\mu(x)$ have the disadvantage of over-describing the system in the sense that different values of $a_\mu(x)$ can correspond to exactly the same physical conditions. Indeed, if we replace $a_\mu(x)$ in (1.1.4) by $a_\mu(x) + \partial_\mu\lambda(x)$ for any function λ of the space-time point x, we will still obtain the same value for the phase difference, and hence the same diffraction pattern on the screen in the Aharonov–Bohm experiment. This means, of course, that the gauge potential $a_\mu(x)$, which we have adopted as dynamical variables, are not in fact physically observable quantities.

Indeed, even the phase difference displayed in (1.1.4) is not an observable quantity, for if it were to change by an integral multiple of 2π, the diffraction pattern will still remain unchanged. What was really physically observable in the Aharonov–Bohm experiment was the *Dirac phase factor*:

$$\Phi(C) = \exp ie \oint_C a_\mu(x)\, dx^\mu \tag{1.3.2}$$

for the line integral in the exponent being taken over a closed curve C such as $\Gamma_2 - \Gamma_1$ in Figure 1.3. Like the field tensor $f_{\mu\nu}$, Φ is gauge invariant and does not suffer from the gauge ambiguity intrinsic in a_μ, but unlike $f_{\mu\nu}$, it gives correctly the phase effect of the electron wave function. Moreover, a change by any integral multiple of 2π in the phase difference of (1.1.4) leaves Φ with the same value. Its description of the physical phenomena is thus exactly right, since given this phase factor, one can specify unambiguously the diffraction pattern on the screen, while no two different values for the phase factor can correspond to the same diffraction pattern. In other words, in contrast to both $f_{\mu\nu}$ and a_μ, the phase factor Φ neither under-describes nor over-describes the dynamical system under consideration.

The situation for nonabelian Yang–Mills theories is similar. The gauge potential A_μ, because of its inherent gauge ambiguity, over-describes the physical system. The field tensor $F_{\mu\nu}$ defined in (1.2.12) under-describes it, this time actually even in the classical theory, as first pointed out by Wu and Yang in 1975; explicit examples are known in which several gauge inequivalent potentials A_μ correspond to exactly the same field tensor $F_{\mu\nu}$. In contrast, the path-ordered phase factor over a closed loop C:

$$\Phi(C) = P \exp ig \oint_C A_\mu(x)\, dx^\mu, \qquad (1.3.3)$$

analogous to (1.3.2) of the abelian theory and sometimes known as the *Wilson loop* in this context, has the requisite properties for describing the physics. Although as it stands, equation (1.3.3), according to (1.2.17), is still gauge dependent, transforming covariantly by a gauge rotation at the starting point (or end-point) of the loop C, this gauge dependence can be removed by taking the trace of $\Phi(C)$. Alternatively, one can restrict consideration only to loops starting from a fixed point P_0, in which case the gauge transformation of $\Phi(C)$ becomes x-independent, occurring only at P_0, and is then easily handled. As will be seen in Chapter 4, these restricted loops are already adequate for a full description of the theory, since the phase factor $\Phi(C)$ for any C is expressible as a product of such factors only of loops starting from and ending at P_0. In the language of Section 3.4 this phase factor is the *holonomy* of the connection A.

The reason why the phase factor Φ succeeds where the field tensor fails is not far to seek. In a sense, Φ is basically just the global version of $F_{\mu\nu}$ in that $\Phi(C)$ specifies parallel phase transport over any loop C including those of finite size, while $F_{\mu\nu}$ specifies the same but only over infinitesimal ones as shown in Figure 1.2 above. Where parallel transport over all finite loops can be built up from that over infinitesimal ones, the two descriptions would be equivalent, but there are situations where this is not the case. The Aharonov–Bohm experiment is already an example. There, the field tensor $f_{\mu\nu}$ being zero in the whole region traversed by the electrons, one may be tempted to deduce that the gauge potential a_μ could be put identically zero everywhere in that region. The conclusion would indeed have been correct if the region under consideration were simply-connected so that every closed curve in it can be continuously shrunk to a point, since then, by Stokes' theorem, the line integral on the right-hand side of (1.1.4) would vanish for all closed curves in the region and the gauge potential could be consistently put to zero everywhere. The reason why this argument fails is of course the fact that the region where $f_{\mu\nu}$ vanishes has a hole in it, representing the magnetized iron core, so that not all closed curves in the region need have a vanishing line integral. In other words, the physical conclusion one drew there depended crucially on the overall 'global' properties of our spatial region, and in this situation one needs the phase factor Φ in (1.3.2), not just the field tensor.

Therefore, apart from being aesthetically the 'correct' variable for describing a gauge theory, the phase factor Φ is expected to be of actual practical value in general when dealing with global properties of gauge fields.

1.4 Specification of Gauge Groups

We have already seen above in the example of the Aharonov–Bohm experiment how global properties of spatial regions can affect physics in local gauge theories. So also situations can arise where the global properties of the gauge group become physically significant. It is necessary therefore to specify in any given theory what exactly is meant by the gauge group.[2] As outlined in Sections 1.1 and 1.2 above, local gauge theories are formulated initially in terms of infinitesimal 'phase' changes which are given by the gauge Lie algebra. When repeated, these infinitesimal 'phase' changes lead to finite changes in the 'phase' represented by exponentials of Lie algebra elements which are then elements of the gauge group. However, this statement by itself does not tell us whether the group elements so generated are all to be regarded as physically distinct. To settle the question, we shall need to examine the given theory in more detail.

Consider first again the simplest example of just a single particle of charge e moving in an electromagnetic field. Under a phase rotation, the wave function ψ of the particle will transform as:

$$\psi \longrightarrow \psi' = e^{ie\alpha}\psi, \qquad (1.4.1)$$

---

[2]In physics it is usual to call the (usually) finite-dimensional symmetry group G the *gauge group*, and the infinite-dimensional group of maps from G to space-time the *group of gauge transformations*. In more mathematical context they are respectively called the *structure group* and the *gauge group*. In this book, we follow the first, i.e. the physicist's, convention.

where, as ascertained above, the factor $\exp ie\alpha$ may be regarded as an element of the gauge group. However, not all the factors for different values of α are to be regarded as physically distinct, since any two values of α differing by $2\pi n/e$ for integral n will obviously give the same value for ψ' and hence be physically indistinguishable. We conclude therefore that the elements of the gauge group are to be labelled, not by α taking all values from $-\infty$ to $+\infty$, but by α only in the range $[0, 2\pi/e]$ with the two end-points identified. In other words, we say that the gauge group in this theory is the *compact* $U(1)$ group with the topology of a circle, and not the *non-compact* group with the topology of the real line.

Suppose next we are told to consider the theory of N charged particles with charges $e_1, e_2, ..., e_N$ moving again in an electromagnetic field. Under a phase rotation, the wave function ψ_r of the rth particle will transform thus:

$$\psi \longrightarrow \psi'_r = e^{ie_r\alpha}\,\psi_r. \tag{1.4.2}$$

The answer to the question of what values of α are to be regarded as yielding distinct group elements now depends on whether the values of the charges $\{e_r\}$ are rational with respect to one another. If the ratio between any pair e_r/e_s is a rational number, i.e. of the form p/q where p and q are integers, then there must exist some value e such that all charges e_r can be written as:

$$e_r = n_r e, \tag{1.4.3}$$

with integral n_r. In that case, all those factors in (1.4.2) with values of α differing by $2\pi n/e$ will give the same $\{\psi'_r\}$ and have indistinguishable physical effects. The gauge group then will again be the compact $U(1)$ group. On the other hand, if there is at least one pair of charges for which the ratio e_r/e_s is irrational, then all values of α in the range from $-\infty$ to $+\infty$ have to be regarded as distinct, and the gauge group becomes non-compact. The term *compact electrodynamics* is thus synonymous with charge quantization. If charge is quantized as in (1.4.3), the gauge group is compact; otherwise, it is non-compact.

One sees from the above example that in order to specify the gauge group, knowing merely the gauge algebra is not enough; one needs to examine the physical content of the theory as to what particles or fields it contains. Let us illustrate the point with another, this time nonabelian, example, say chromodynamics. Suppose we consider first just the pure Yang–Mills theory, namely the theory with only colour gluons and no quarks. The gauge potential representing the gluon field transforms under a gauge transformation as in (1.2.7) above where for chromodynamics S is an $SU(3)$ matrix, i.e. a 3×3 unitary matrix with unit determinant. The question then arises whether all such matrices are to be regarded as distinct elements of the gauge group.

To answer this question, we observe first that the following matrices are unitary and have unit determinant, and thus form a cyclic subgroup, say $\mathbb{Z}_3$, of $SU(3)$:

$$\zeta_r = \exp i2\pi r/3, \quad r = 0, 1, 2. \tag{1.4.4}$$

Besides, being scalar multiples of the identity matrix, they commute with all elements of the $\mathfrak{su}(3)$ algebra, and hence also with $A_\mu(x)$. Thus, given any gauge transformation parametrized by an $S(x)$ belonging to $SU(3)$, we may consider in conjunction two other transformations parametrized by $S\zeta_r, r = 1, 2$ which, ζ_r being x-independent and commuting with $A_\mu(x)$, will have, on substitution into (1.2.7), the same effect on $A_\mu(x)$ as S. And since there are by definition no other fields than $A_\mu(x)$ in the pure Yang–Mills theory, it follows that the latter two transformations will be physically indistinguishable from the first. We conclude therefore that the gauge group in this theory is not $SU(3)$ but a group obtained from $SU(3)$ by identifying, for all γ belonging to $SU(3)$, triplets of the form: $\gamma, \gamma\zeta_1, \gamma\zeta_2$, a group which we may denote as $SU(3)/\mathbb{Z}_3$.

The conclusion reached above can readily be generalized to Yang–Mills theories with gauge algebra $\mathfrak{su}(N)$ for any N. For the pure Yang–Mills theory containing only the gauge potential A_μ, the gauge group is not $SU(N)$, i.e. the group of all $N \times N$ unitary matrices with unit determinant, but the group obtained from $SU(N)$ by identifying all N-tuples of the form $\gamma\zeta_r, r = 0, 1, ..., N-1$, for

$$\zeta_r = \exp i2\pi r/N, \tag{1.4.5}$$

namely the group $SU(N)/\mathbb{Z}_N$, where $\mathbb{Z}_N$ is the cyclic subgroup of order N which commutes with all elements of $SU(N)$. In particular, for $N = 2$, the pure Yang–Mills theory has gauge group $SU(2)/\mathbb{Z}_2 = SO(3)$, a simple example which we shall have occasion later to study in some detail.

The gauge group will in general be different when the theory contains other particles or fields. For example, in chromodynamics containing both quarks and gluons, the gauge group is the full $SU(3)$ and not $SU(3)/\mathbb{Z}_3$, for although the elements ζ_r in (1.4.4) remain indistinguishable in their effect on A_μ, they operate differently on the quark fields:

$$\zeta_r \psi_q = \exp i2\pi r/3 \, \psi_q \neq \psi_q, \quad r \neq 0, \tag{1.4.6}$$

and are therefore physically distinct.

As yet another example, it is worthwhile to consider the standard electroweak theory, the gauge group for which is often referred to in the literature as $SU(2) \times U(1)$, but this is, strictly speaking, incorrect. By definition, the direct product $A \times B$ of two groups consists of the couples (a, b) where $a \in A$ and

$b \in B$, and different couples are to be regarded as distinct elements of $A \times B$.
Hence, the gauge group for the standard electroweak theory is $SU(2) \times U(1)$
only if all couples (f,y) for all $f \in SU(2)_f$ and $y \in U(1)_Y$ are physically
distinguishable. However, as the theory stands at present, the physical fields
are either:

(i) $SU(2)$ doublets with half-integral hypercharges, e.g. $(\nu, e)_L$ with $Y =
1/2$; (ϕ^+, ϕ^0) Higgs with $Y = -1/2$,
or else:

(ii) $SU(2)$ singlets or triplets with integral hypercharges, e.g. $(e)_R$ with
$Y = 1$, gauge bosons B_μ, A_μ^i with $Y = 0$.

Hence, if we take:

$$\tilde{f} = \exp i2\pi T_3 \quad \in SU(2)_f, \tag{1.4.7}$$

$$\tilde{y} = \exp i2\pi Y \quad \in U(1)_Y, \tag{1.4.8}$$

then, since all fields in the theory have $T_3 + Y$ integral,

$$(\tilde{f}, \tilde{y})\psi = \exp i2\pi(T_3 + Y)\,\psi = \psi, \tag{1.4.9}$$

or that the couple $(\tilde{f}, \tilde{y})$ has the same effect as the identity $(1,1)$ on all fields
in the theory. This means that the couple $(f\tilde{f}, y\tilde{y})$ will be physically indistinguishable from the couple (f,y) for any $f \in SU(2)_f$ and $y \in U(1)_Y$. Therefore,
the gauge group for the standard electroweak theory is not $SU(2) \times U(1)$, but
a group obtained from $SU(2) \times U(1)$ by identifying $(f\tilde{f}, y\tilde{y})$ with (f,y) for all
f,y, which we may, if we like, denote by $[SU(2) \times U(1)]/\mathbb{Z}_2$, but is in fact just
identical to the group $U(2)$ of 2×2 unitary matrices. Notice that in (1.4.9)
above we have used the 2–1 map from $SU(2) \times U(1)$ to $U(2)$ which is explicitly
given by multiplication: $(f,y) \to fy$, where the element $y \in U(1)$ is written as
a 2×2 scalar matrix so that it can be multiplied with the matrix $f \in SU(2)$
to give a matrix with arbitrary determinant in $U(2)$. This also makes it clear
mathematically why in the case of (i) above the couple $(f\tilde{f}, y\tilde{y}) = (-f, -y)$ is
identified with the couple (f,y) when considered as elements of $U(2)$.

A similar analysis applied to the *standard model* of electroweak theory
coupled to chromodynamics containing the usual medley of quarks, leptons,
gluons, electroweak gauge bosons and Higgs, reveal that the gauge group is
strictly speaking not $SU(3)_c \times SU(2)_f \times U(1)_Y$ as usually stated in the literature
but a group obtained from it by identifying the following sextuplets of elements:

$$\begin{aligned}(c, f, y) &= (cc_1, f, yy_1) = (cc_2, f, yy_2)\\ &= (c, f\tilde{f}, y\tilde{y}) = (cc_1, f\tilde{f}, y\hat{y}y_1) = (cc_2, f\tilde{f}, y\hat{y}y_2)\end{aligned} \tag{1.4.10}$$

for any $c \in SU(3)_c$, $f \in SU(2)_f$, and $y \in U(1)_Y$, where

$$T = T_3 - Q \tag{1.4.11}$$

$$c_r = \exp\frac{i2\pi r}{\sqrt{3}}\lambda_8, \quad r = 1, 2, \tag{1.4.12}$$

$$y_r = \exp i4\pi rY, \quad r = 1, 2, \tag{1.4.13}$$

$$\hat{y} = \exp i6\pi Y, \tag{1.4.14}$$

and $\tilde{f}$ is as defined in (1.4.7) and above. We may denote the group by $[SU(3)_c \times SU(2)_f \times U(1)_Y]/\mathbb{Z}_6$, although the notation is somewhat ambiguous. Notice that since Y is quantized here in $1/6$ units, the period of $U(1)_Y$ is 12π.

Of course, it is not excluded that future experiments may force us to change our theory to include particles or fields belonging to other representations; then the gauge group will have to be redefined accordingly. For example, if particles with $T + Y$ taking half-integral values were discovered, the gauge group for the electroweak theory will have to be $SU(2)_f \times U(1)_Y$ and not $U(2)$ as at present.

Although the discussion above may seem at first sight a little pedantic, it will be seen later that a proper specification of the gauge group is essential, for example, in the study of monopoles.

Chapter 2

Monopoles

2.1 Magnetic Charge as Topological Obstruction

One of the most intriguing features of local gauge theories, first noted by Dirac, is the natural occurrence in them of monopoles as topological obstructions. To see this, consider a one-parameter family of closed loops, say $\{C_t\}$, labelled by a parameter t running from 0 to 2π, enveloping a closed surface Σ in the manner depicted in Figure 2.1. At $t = 0$, C_t is just the point P_0. As t increases, C_t gradually expands, loops over the surface Σ, then becomes smaller again until at $t = 2\pi$, it shrinks back to the point P_0. For each loop C_t, we can evaluate the phase factor $\Phi(C_t)$, using (1.3.3), obtaining an element of the gauge group for each value of t. At $t = 0$ and 2π, the loop shrinks to a point, which means that the line integral in the exponent vanishes and one obtains for $\Phi(C_0)$ and $\Phi(C_{2\pi})$ the identity element. Hence, as t varies continuously from 0 to 2π, $\Phi(C_t)$ traces out a continuous curve, say Γ, in the gauge group, beginning and ending at the identity element. The question then is the following. Suppose we continuously deform the surface Σ, reducing it eventually to a point, will the curve Γ in the gauge group also contract then to a point? If not, then it would appear that there is something inside the surface Σ which is obstructing the contraction. That something if it exists would be inherent in the topological structure of the gauge field and could for that reason be termed a *topological obstruction*. The above procedure is rather analogous to that of a cowboy playing with a lasso. By throwing $\{C_t\}$ over Σ, we have enveloped with it a region in space. Then, by squeezing the surface Σ to a point, continuously so that nothing can escape, we shall discover whether we have caught anything of interest inside. We shall see that, depending on the gauge group, the answer can in some cases be positive, and the captive

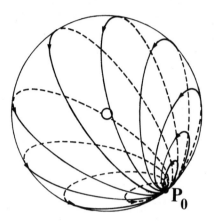

Figure 2.1: Looping over a monopole.

will be the elusive monopole.

Under what conditions on the gauge group will our catch be nontrivial? As asserted above, given a one-parameter family of closed loops $\{C_t\}$ enveloping a surface Σ, we have a closed curve Γ in the gauge group G. As Σ is continuously deformed in space, so also will Γ be in G. The question then is whether, by continuously deforming Σ to a point in space, Γ can also be deformed to a point in G. The answer will depend on the topology of G. If G is simply-connected, as illustrated in Figure 2.2a, then any closed curve Γ in it is continuously

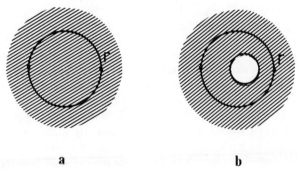

a   **b**

Figure 2.2: Simply-connected and non-simply-connected regions.

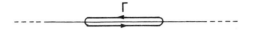

Figure 2.3: A closed curve on the real line.

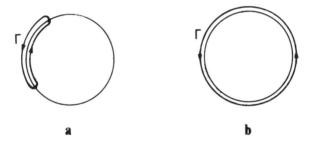

a   b

Figure 2.4: Closed curves on the circle $U(1)$.

deformable to a point, but if G has holes in it, i.e. multiply-connected, as illustrated in Figure 2.2b, then some closed curves in it can never be shrunk continuously to a point. In the latter case, we shall have monopoles.

Let us again take first electromagnetism as example. If we are dealing with *non-compact electrodynamics*, as specified in Section 1.4, the gauge group is the real line; then any closed curve Γ in it will be of the form shown in Figure 2.3 and can always be deformed continuously to a point. On the other hand, if we are dealing with *compact electrodynamics*, the gauge group is $U(1)$, which has the topology of a circle. In that case, some closed curves indeed can still be shrunk continuously to points, as for example that shown in Figure 2.4a, but there are others which cannot be so deformed, such as the example depicted in Figure 2.4b which winds once around the circle representing the gauge group. Hence, appending this information to our considerations above, we conclude that monopoles exist in electrodynamics if and only if the gauge group is compact, or equivalently, if and only if charge is quantized, which was the astounding result first brilliantly derived by Dirac in 1931.

Our contention above was that the case of Figure 2.4b should correspond to a *monopole*, in this case a magnetic charge. Let us see now whether such a claim can be maintained. We recall that 'charge' in electromagnetism is conventionally defined as a source of flux; thus, in the case of a magnetic charge, one of magnetic flux. To show that there is a magnetic charge inside the surface Σ, we shall need to measure the total magnetic flux emerging from that surface. We start by evaluating the line integral in the exponent of the phase factor in (1.3.2) for each C_t, namely:

$$\alpha_t = e \oint_{C_t} a_\mu(x)\, dx^\mu, \qquad (2.1.1)$$

which is the angle giving the position of the point at t of the curve Γ on the circle representing the gauge group in Figure 2.4. The direct evaluation of α_t in terms of a_μ is not as straightforward as it may seem, since, as we shall see later, in the presence of a monopole, a_μ is bound to be singular on any surface surrounding that monopole. However, we can instead evaluate α_t in (2.1.1) by Stokes' theorem, giving:

$$\alpha_t = -e \iint_{\Sigma_t} f_{\mu\nu}(x)\, d\sigma^{\mu\nu}, \qquad (2.1.2)$$

where Σ_t may be taken to be the surface swept out by $\{C_{t'}\}$, for $t' = 0 \to t$. In other words, α_t is e times the magnetic flux traversing the surface Σ_t bounded by C_t. As $t \to 2\pi$, C_t, by construction, shrinks back to the point P_0, so that Σ_t becomes the whole surface Σ depicted in Figure 2.1, and α_t becomes e times the total amount of magnetic flux emerging from the closed surface Σ. Now for the case depicted in Figure 2.4a, the phase α_t starts at 0 at the origin, and returns to the origin such that at $t = 2\pi$, α_t is again zero. The total magnetic flux through Σ is thus zero and we have not captured anything. However, for the case depicted in Figure 2.4b, although α_t starts again at zero at $t = 0$ at the origin, it returns there at $t = 2\pi$ only after winding around the circle once, which means that at $t = 2\pi$, $\alpha_t = 2\pi$ instead. Hence, the total magnetic flux through Σ is $2\pi/e$ in this case. This is exactly the value we would obtain if there was a magnetic monopole inside Σ with magnetic charge $\tilde{e} = 1/(2e)$. So we have shown that we have indeed captured by $\{C_t\}$ a magnetic charge, or that in electromagnetism the definition above of a magnetic monopole as a topological obstruction is equivalent to its conventional definition as a source of magnetic flux.

What other values can the monopole charge take? We recall that Γ is required to be a closed curve in the gauge group G beginning and ending at the identity. In the present case where G is a circle, Γ must therefore wind around the circle an integral number of times. Suppose we adopt the convention that the integer, say n, is positive (negative) if Γ winds around the circle in the counter-clockwise (clockwise) direction. Then, by repeating the argument in the preceding paragraph, one easily concludes that if Γ winds around the circle n times, the total magnetic flux emerging from Σ is $2n\pi/e$, or that the magnetic charge enveloped by $\{C_t\}$ is:

$$\tilde{e} = n/(2e). \tag{2.1.3}$$

Moreover, one can conclude that the magnetic charge can take no value other than these, namely integral (positive or negative) multiples of $1/(2e)$. This result is the celebrated *Dirac quantization condition* first obtained by Dirac in 1931 via some very ingenious arguments.

In addition, one can deduce that this quantized magnetic charge is a conserved quantity both in classical and in quantum dynamics. That this is the case can be seen as follows. Clearly, by continuous deformations of the closed curve Γ, one cannot change its winding number around the circle representing G from one of its discrete values n to another $n' \neq n$. Consider first classical dynamics; here the dynamical system evolves in time by a continuous variation of its dynamical variables, which, for electromagnetism, may be taken to be the gauge potential $a_\mu(x)$. As $a_\mu(x)$ varies in time, so will the phase factor $\Phi(C_t)$ and hence also the closed curve Γ in G. However, if $a_\mu(x)$ is allowed to vary only continuously, the corresponding continuous variations in Γ can never change its winding number n around G, so that the magnetic charge $\tilde{e}$ in (2.1.3) must also remain the same. That the identical conclusion holds even for the quantum theory is slightly more subtle. Time evolution here is given in terms of the transition matrix whose element between an initial state $|i\rangle$ and a final state $|f\rangle$ can symbolically be represented as a Feynman integral:

$$\langle f|i\rangle = \int_i^f \delta a \, e^{-i\mathcal{A}[a]}, \tag{2.1.4}$$

where $\mathcal{A}$ is the action, which is some functional of the gauge potential $a = \{a_\mu(x)\}$. The integral in (2.1.4) is to be taken over all continuous 'paths' linking $|i\rangle$ to $|f\rangle$, or in other words, over all continuous functions $a_\mu(x)$ which at $t = t_i$ take some values specifying the initial state $|i\rangle$ and at $t = t_f$, some other values specifying the final state $|f\rangle$. Suppose now we take $|i\rangle$ and $|f\rangle$ to be states with two distinct values of the magnetic charge in (2.1.3); then, according to our observation above, there can be no continuous 'path' in a linking one state to the other. Hence, the integration region in (2.1.4) will vanish and so will the transition matrix element $\langle f|i\rangle$, so that the transition probability from $|i\rangle$ to $|f\rangle$ will be zero. Again, therefore, the magnetic charge $\tilde{e}$ will have to be conserved.

We notice that the conservation law for the magnetic charge was derived above without any reference to the specific form of the action $\mathcal{A}$ or to the detailed formulation of the dynamics; it arose as a consequence merely of the primitive concepts of gauge invariance and of continuity. It is thus something very intrinsic, and is, in a sense, even more fundamental conceptually than the conservation laws deduced from the Noether theorem which arose from a detailed application of the Action Principle. We shall have occasion to discuss this point further in the future.

2.2   Nonabelian Monopoles

One sees from the above section that merely by applying the basic gauge theory concepts outlined in the preceding chapter, we were able to derive for compact electrodynamics the important result not only that the theory admits the existence of monopoles, but that their charges must be quantized and conserved. It is clearly of interest to ask whether similar conclusions can be drawn when considering other Yang–Mills fields. In general terms, the question hinges on whether closed curves Γ on the gauge group G can all be shrunk continuously to a point. Those which cannot be so eliminated will correspond to some sort of topological charge. The values that this topological charge can take are labelled by the distinct classes of curves Γ which cannot be continuously deformed into one another. Mathematically, they are known as *homotopy classes* which we shall define more properly later in Chapter 3. They are an intrinsic property of the gauge group G. By repeating then the arguments given above for electrodynamics, one sees that the topological charge corresponding to one homotopy class cannot evolve in time into one corresponding to another homotopy class and is thus again a conserved quantity. Besides, the homotopy classes being by definition discrete, these charges are also in a sense quantized. In view of the similarity these new charges have with magnetic monopoles in the abelian theory, we shall call them *nonabelian monopoles*.[1]

Superficially, it might appear that, instead of the somewhat abstract procedure above, a more obvious way of generalizing the magnetic monopole to a

---

[1] To avoid confusion, it is probably worth pointing out already at this stage that the term *nonabelian monopole* is often applied in the literature also to another completely different object. In certain nonabelian theories with spontaneous symmetry breaking, there may exist soliton solutions carrying monopole charges. The first and simplest examples of these were given in the famous papers of 't Hooft and Polyakov of 1974, in which solitons in an $SO(3)$ theory were found carrying magnetic charges. These are sometimes referred to also, inappropriately we think, as 'nonabelian monopoles', but in the language adopted here they are just $U(1)$-monopoles embedded in a theory with a larger gauge group. In mathematical language, to be explained lated in Chapter 3, these magnetically charged solitons in the $SO(3)$ theory are nontrivial (but abelian) $U(1)$-subbundles of a nonabelian but trivial $SO(3)$-bundle. On the other hand, the nonabelian monopoles just defined in the text above are genuine nontrivial nonabelian bundles. Some discussion on the relationship between the two can be found later in Sections 2.6 and 3.4.

nonabelian theory would be to generalize the conventional concept of a source of magnetic flux. When one attempts to do so, however, one is soon convinced that this is not the case. Indeed, to ascertain whether there is a magnetic monopole in a given region of space, the conventional approach is to envelop the region by a closed surface and measure the total magnetic flux emerging from it. Equivalently, using Gauss' theorem, the total flux can be expressed as a volume integral of the divergence of the magnetic field $\mathbf{H}$ over the region enclosed. Except when there is a monopole inside, the measured flux will be zero because

$$\operatorname{div} \mathbf{H} = 0. \tag{2.2.1}$$

This obtains because the electromagnetic field $f_{\mu\nu}$ is a gauge field, meaning that it can be derived from a gauge potential a_μ according to (1.1.6), and thus automatically satisfies the Bianchi identity:

$$\partial_\lambda f_{\mu\nu} + \partial_\mu f_{\nu\lambda} + \partial_\nu f_{\lambda\mu} = 0. \tag{2.2.2}$$

This can also be written as:

$$\partial^\nu \,{}^*f_{\mu\nu} = 0, \tag{2.2.3}$$

where ${}^*f_{\mu\nu}$, known as the *dual field tensor*, is defined as:

$$\,{}^*f_{\mu\nu} = -\tfrac{1}{2}\epsilon_{\mu\nu\rho\sigma} f^{\rho\sigma} \tag{2.2.4}$$

with $\epsilon_{\mu\nu\rho\sigma}$ being the totally antisymmetric tensor. And (2.2.1) is just the 0th component of (2.2.3), since by definition $\mathbf{H}$ is:

$$H_i = -\tfrac{1}{2}\epsilon_{0ijk} f^{jk} = {}^*f_{0i}. \tag{2.2.5}$$

Suppose we try now to repeat the argument for nonabelian fields. We find that for $F_{\mu\nu}$ defined as in (1.2.12), one still has a Bianchi identity:

$$D_\lambda F_{\mu\nu} + D_\mu F_{\nu\lambda} + D_\nu F_{\lambda\mu} = 0, \tag{2.2.6}$$

or,

$$D^\nu \,{}^*F_{\mu\nu} = 0, \tag{2.2.7}$$

with the dual field ${}^*F_{\mu\nu}$ defined similarly to (2.2.4), namely

$$\,{}^*F_{\mu\nu} = -\tfrac{1}{2}\epsilon_{\mu\nu\rho\sigma} F^{\rho\sigma}, \tag{2.2.8}$$

where the derivative involved is the *covariant derivative*:

$$D_\mu = \partial_\mu - ig[A_\mu, \ ]. \tag{2.2.9}$$

However, the total 'flux' emerging from any closed surface, if defined in naive analogy with the abelian case again just as:

$$\text{'flux'} = -\iint F_{\mu\nu}\, d\sigma^{\mu\nu}, \qquad (2.2.10)$$

would still be given as an integral of the ordinary divergence $\partial^\nu\, {}^*F_{\mu\nu}$, which, according to (2.2.7) need not be zero even in the absence of a monopole inside the surface. It cannot therefore be used to characterize a monopole. Attempts have of course been made to generalize the concept of 'flux' obtained from electromagnetism to nonabelian theories in a more sophisticated fashion. They all eventually lead back to the phase factor Φ in (1.3.3), and hence to a conception of the nonabelian monopole equivalent to that promulgated above.

Returning then to our previous definition of the nonabelian monopole charge, let us examine a few specific examples. We stressed above that in electrodynamics, the correct answers to the questions of whether monopoles exist, and if they do, of what charges they may possess, are crucially dependent on a proper specification of the gauge group. One expects the same to be true in nonabelian theories. Take in particular theories with gauge algebra $\mathfrak{su}(N)$. It was noted in Section 1.4 that if we deal with the pure Yang–Mills theory containing only gauge bosons, the gauge group is $SU(N)/\mathbb{Z}_N$, but if the theory contains other fields belonging to the fundamental representation, such as quarks in QCD, then the gauge group is $SU(N)$. We shall now show that whereas monopole charges are admitted by the former theory, no monopole charge exists in the latter theory.

It is sufficient to illustrate with the simplest example, namely the theory with the gauge algebra $\mathfrak{su}(2)$. The group $SU(2)$ has the topology of a 3-dimensional hyperspherical surface S^3. That this is the case can be seen as follows. By definition, $SU(2)$ is the group of 2×2 unitary matrices with unit determinant. Imposing then on the general 2×2 matrix with complex elements:

$$A = \begin{pmatrix} a & b \\ c & d \end{pmatrix} \qquad (2.2.11)$$

the condition that it be unitary and have a unit determinant, it is easily seen that a, b, c, d in (2.2.11) must satisfy:

$$a = \bar{d}, \quad b = -\bar{c}, \quad a\bar{a} + b\bar{b} = 1. \qquad (2.2.12)$$

The group elements can therefore be parametrized by 4 real parameters, say x, y, z, and w, with $a = x + iy$ and $b = z + iw$, satisfying:

$$x^2 + y^2 + z^2 + w^2 = 1, \qquad (2.2.13)$$

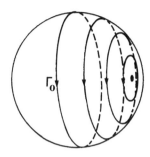

Figure 2.5: Closed curves in SU(2).

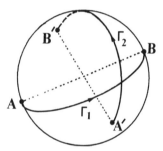

Figure 2.6: Closed curves on SO(3).

which is the equation of a 3-sphere in 4-space with x, y, z, w as Cartesian coordinates. That being the case, it is readily seen that a theory with gauge group $SU(2)$ cannot admit monopoles in the sense defined above since any closed curve Γ_0 on S^3 can be continuously deformed to a point, as illustrated in Figure 2.5.

On the other hand, if we are dealing with the pure Yang–Mills theory, the gauge group is no longer $SU(2)$ but $SU(2)/\mathbb{Z}_2 \cong SO(3)$, which is obtained from $SU(2)$ by identifying pairs of elements with opposite signs. In that case, the curve Γ_1 on S^3 depicted in Figure 2.6 is also a closed curve on $SO(3)$ since the two *antipodal points* A and B in $SU(2) \cong S^3$ are to be regarded as the same point in $SO(3)$. Now, this curve cannot be shrunk to a point by any continuous deformation and should thus, according to the reckoning above, correspond to a monopole. Next, let us ask what values the monopole charge can take. One sees that by continuously deforming Γ_1, one can obtain 'closed'

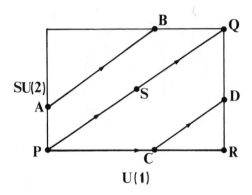

Figure 2.7: Schematic representation of the group   U(2).

curves Γ_1' in $SO(3)$ joining any other pairs of antipodal points A' and B' on S^3, as illustrated in Figure 2.6. This means that all these curves belong to the same homotopy class of $SO(3)$ and correspond to the same monopole charge as that of Γ_1 above. Further, if one combines two such monopole charges, one obtains a curve which winds around $SO(3)$ twice, namely around a complete circuit on $SU(2)$. The result is a curve similar to Γ_0 in Figure 2.5, which can be deformed continuously to a point. This means that a monopole here is the same as an anti-monopole since two monopoles combine to give total charge zero. The final conclusion is therefore that the monopole charge in this theory can take values only in $\mathbb{Z}_2$, i.e. integers modulo 2. Equivalently, we can denote the monopole charge by a sign $\pm$, where $+$ corresponds to the vacuum, and $-$ to a monopole.

The extension of the above analysis to $\mathfrak{su}(N)$ theories is straightforward. Theories with gauge group $SU(N)$ have no monopoles, but the pure gauge theories with gauge group $SU(N)/\mathbb{Z}_N$ will have monopoles whose charges take values in $\mathbb{Z}_N$. Thus, in particular, full QCD containing both gluons and colour triplet quarks has no monopoles while pure QCD with only gluons admit monopoles with charges labelled by a 'triality', i.e. integers modulo 3, or equivalently the cube roots of unity: $\zeta_r = \exp i2\pi r/3$, $r = 0, 1, 2$.

As a further example, consider the standard electroweak theory. In this case, the gauge group was identified in Section 1.4 as $U(2)$ or $[SU(2) \times U(1)]/\mathbb{Z}_2$ by which we meant that $U(2)$ can be obtained from $SU(2) \times U(1)$ by identifying certain pairs of elements as explained in the paragraph after (1.4.9). The group $SU(2) \times U(1)$ itself consists of couples of elements from respectively the groups $SU(2)$ and $U(1)$, which we may represent symbollically as points inside the rectangle in Figure 2.7 where the vertical axis represents the group

$SU(2)$ and the horizontal axis the group $U(1)$, and the parallel edges of the rectangle are understood to be identified so as to make the rectangle into a 'hyper-torus'. Hence the diagonal PQ in Figure 2.7 represents a closed curve in the group $SU(2) \times U(1)$, since P and Q represent the same element in the group; similarly, $ABCD$ is also a closed curve in $SU(2) \times U(1)$. Both these curves wind once around each of the two subgroups $SU(2)$ and $U(1)$. Now $SU(2)$ being simply-connected, winding around $SU(2)$ once is homotopically the same as not winding around at all. In other words, a curve like PQ can always be deformed continuously in $SU(2) \times U(1)$ into another curve like PR which winds once only around the group $U(1)$ but not around the group $SU(2)$. On the other hand, as explained above in Figure 2.4b, the winding number of a curve around $U(1)$ cannot be changed by any continuous deformation. Curves winding several times around $SU(2) \times U(1)$ can be similarly considered. Hence we conclude that there are nontrivial homotopy classes of closed curves in $SU(2) \times U(1)$ which are labelled just by the homotopy classes of $U(1)$. Equivalently, in physical terms, this means that in a theory with gauge group $SU(2) \times U(1)$ there can be monopoles with charges taking integral values. They will carry a $U(1)$-monopole charge corresponding to the winding number around $U(1)$, but of course no $SU(2)$-monopole charge since none exists.

What we are interested in, however, is not a theory with gauge group $SU(2) \times U(1)$, but the standard electroweak theory with gauge group $U(2)$. In that case, according to the analysis in Section 1.4, the point S in Figure 2.7, representing the couple $(\tilde{f}, \tilde{y})$ in $SU(2) \times U(1)$ for $\tilde{f}$ and $\tilde{y}$ defined as in (1.4.7) and (1.4.8), is to be regarded as identical to $P = (1, 1)$, so that PS is already a closed curve in $U(2)$. Now PS winds only half-way around each of $SU(2)$ and $U(1)$, and should, according to our previous reckoning, correspond to $SO(3)$-monopole charge $-$, and a $U(1)$-monopole charge only $\frac{1}{2}$ of that represented by PQ or PR above. As in the pure $SO(3)$ theory, the $SO(3)$-monopole charge $-$ here cannot be eliminated by any continuous deformation. Further, the above arguments can readily be extended to other curves winding several times around $U(2)$. The conclusion is then that the standard electroweak theory admits monopoles with charges labelled by integers. A charge labelled by the integer n may be regarded as carrying simultaneously an $SO(3)$-monopole charge $(-1)^n$ and a $U(1)$-monopole charge of $n/(4g')$. Notice that the $U(1)$-monopole charge is here quantized in units of $1/(4g')$ instead of the usual unit $1/(2g')$ in (2.1.3) for the simple $U(1)$ theory.

A similar analysis can be applied to the standard model combining the electroweak theory with chromodynamics, which, as specified in Section 1.4, has a gauge group denoted there by $[SU(3) \times SU(2) \times U(1)]/\mathbb{Z}_6$. The conclusion is again that monopoles are admissible in this theory, and that they have charges labelled by integers. A monopole in this theory with charge labelled by n may be regarded as carrying simultaneously (i) an $[SU(3)/\mathbb{Z}_3]$-monopole charge of $\exp i2\pi n/3$, (ii) an $SO(3)$-monopole charge of $(-1)^n$, and (iii) a $U(1)$-monopole charge of $n/(6g')$.

From the examples considered above, it is now clear that the question of what monopole charges exist in any given theory has been reduced to a mathematical problem. For a more rigorous and general formulation of the problem together with a selection of the relevant mathematics for solving it, the reader is referred to Chapter 3. As physicists, however, we would want to know more. For example, we would like to know how these abstractly defined charges will behave when brought into interaction with other particles or fields. Now, for electromagnetism, we already have some preconception on what magnetic charges will do in the presence of an electromagnetic field. We need therefore to examine from where these ideas originate in electromagnetism, and then ask whether the same type of reasoning can be applied to make similar assertions about their nonabelian analogues. It will appear later that one can indeed do so, but one will need first to develop some tools. We shall return to reconsider this problem in Chapter 5.

2.3   Patching away the Dirac String

As mentioned in Section 2.1, the electromagnetic potential a_μ in the presence of a monopole must be singular. Indeed, it must be singular at at least one point on every closed surface surrounding a monopole. That this is the case can be seen as follows. Let us calculate the total magnetic flux leaving — say, for simplicity — a sphere. Suppose the potential a_μ is nonsingular everywhere on the sphere. Then, using Stokes' theorem, one can write the flux from the northern hemisphere as:

$$-\iint_N f_{ij}\,d\sigma^{ij} = \oint a_i\,dx^i, \qquad (2.3.1)$$

where the line integral on the right is to be taken along the equator in the counter-clockwise direction when viewed from the north pole. Similarly, the flux from the southern hemisphere is:

$$-\iint_S f_{ij}\,d\sigma^{ij} = \oint a_i\,dx^i, \qquad (2.3.2)$$

where now the line integral along the equator has to be taken in the clockwise direction when viewed from the north pole. Clearly, on adding the two contributions together, one gets zero for the total magnetic flux, in contradiction with the assertion that there is a magnetic monopole inside, which should give a total flux of $4\pi\tilde{e}$ for a monopole of charge $\tilde{e}$. One concludes therefore that the supposition above was wrong, or in other words that a_μ must have a singularity somewhere on the sphere.

The above arguments hold for any sphere surrounding the monopole. Hence, by increasing continuously the radius of the sphere, the singularity on it will trace out a continuous curve. In other words, we have deduced that a monopole must be attached to a whole line of singularities stretching all the way to infinity. Such a line of singularities, first noted by Dirac, is called a *Dirac string*.

For future reference, we give here as example the potential for a static monopole. The magnetic field is in this case:

$$\mathbf{H} = \tilde{e}\mathbf{r}/r^3, \tag{2.3.3}$$

which can be derived from the potential,

$$a_x = \frac{\tilde{e}y}{r(r+z)}, \quad a_y = \frac{-\tilde{e}x}{r(r+z)}, \quad a_z = 0. \tag{2.3.4}$$

This is singular all along the negative z-axis, which is then our Dirac string.

We notice, however, that a system containing only a single static charge is spherically symmetric so that the negative z-axis can have no special significance. Indeed, we could equally well have chosen to derive the magnetic field (2.3.3) from the potential:

$$a'_x = \frac{-\tilde{e}y}{r(r-z)}, \quad a'_y = \frac{\tilde{e}x}{r(r-z)}, \quad a'_z = 0, \tag{2.3.5}$$

instead of (2.3.4), placing thus the Dirac string along the positive z-axis. The difference between (2.3.4) and (2.3.5) is just a gauge transformation of the form (1.1.2):

$$(a'_i - a_i) = \partial_i\alpha, \quad \alpha = 2\tilde{e}\arctan(y/x) = 2\tilde{e}\phi, \tag{2.3.6}$$

as can be checked by direct computation.

From the above example, then, one sees that the Dirac string is actually not a physical singularity at all, but merely a singularity in our representation of the potential in a particular gauge choice. It is similar in nature to the coordinate singularity in the cartographer's zenithal projection of the globe. There, if the north pole is chosen as the zenith, then the coordinates of the north pole are singular, and if the south pole is chosen as the zenith, then the coordinates of the south pole are singular, although there is no actual singularity on the globe itself.

To give a full representation of the globe, the cartographer usually uses two zenithal projections, say, one from the north pole and one from the south pole; then specifies which point in one projection corresponds to which point in the other in the region where the two projections overlap, e.g. by giving the longitudes of points in the equatorial region on both charts. In a very similar fashion, one can give a fully nonsingular representation of the gauge potential a_μ over the whole surface surrounding a monopole provided that one pays the price of using more than one coordinate patch and specifying in the overlap regions of these patches the relation between the potentials defined in the different patches. Clearly, just as for the cartographer's zenithal projections, such *patched representations* of gauge potentials are not unique; apart from convenience, it is immaterial whether one chooses to cover the surface of the sphere by two or more patches, or what size and shape one assigns to each patch.

As an example, let us represent the potential of a static magnetic monopole in this way. Let us cover a sphere enclosing the monopole by two patches, say, the northern and southern hemispheres:

$$(N): \ 0 \le \theta < \pi, \ 0 \le \phi \le 2\pi,$$
$$(S): \ 0 < \theta \le \pi, \ 0 \le \phi \le 2\pi, \tag{2.3.7}$$

overlapping in a band along the equator. In (N) we choose:

$$a_i^{(N)}(x) = a_i(x), \ a_0^{(N)} = 0, \tag{2.3.8}$$

for $a_i(x)$ given in (2.3.4), and in (S):

$$a_i^{(S)}(x) = a_i'(x), \ a_0^{(S)} = 0, \tag{2.3.9}$$

for $a_i'(x)$ given in (2.3.5). In the overlapping region along the equator, $a_\mu^{(N)}(x)$ and $a_\mu^{(S)}(x)$ are related by (2.3.6). Since the negative z-axis is excluded from (N) and the positive z-axis is excluded from (S), both $a_\mu^{(N)}(x)$ and $a_\mu^{(S)}(x)$ are nonsingular in the respective domains in which they are defined. However, although the expressions (2.3.1) and (2.3.2) for the magnetic flux through each hemisphere still apply, they no longer cancel, since the line integral on the right of (2.3.1) now refers to the potential $a_\mu^{(N)}(x)$, while that of (2.3.2) refers to $a_\mu^{(S)}(x)$, and according to (2.3.6):

$$\oint_{\text{equator}} (a_\mu^{(N)} - a_\mu^{(S)}) \, dx^\mu = 4\pi\tilde{e}, \tag{2.3.10}$$

which is exactly what it ought to be for the total flux emerging from the magnetic charge $\tilde{e}$. Notice that although the expressions (2.3.8) and (2.3.9) for the potential are specific to the static monopole in a particular gauge, the *patching condition* (2.3.10) is not, and applies as well to moving monopoles in any gauge patched in the manner (2.3.7), since the monopole charge is the same throughout.

It is instructive at this point to try to visualize what patching of the monopole potential means. We recall that the phase of the wave function ψ of an (electrically) charged particle can range from 0 to 2π and that it is the gauge potential a_μ which specifies what value of this phase at one space-time point should be regarded as being 'parallel' to a given value at a neighbouring point. So we can imagine that to every point x there is attached a circle representing the various values that the phase of ψ can take, and $a_\mu(x)$ is what specifies which point on the circle at $x + dx^\mu$ corresponds to which point on the circle at x. Take now some sphere in space and attach a circle to every point on it. The resulting geometrical construct can be just the direct product of the sphere S^2 with a circle S^1, namely $S^2 \times S^1$, as depicted symbolically in Figure 2.8b', but only in the case when the sphere under consideration does not enclose a monopole. If there is a monopole charge inside, then the topology will have to be more complicated, for although we do not know in general exactly what form the gauge potential will take, we do know that it has to be patched and that the patching condition (2.3.10) has to be satisfied. Hence along the equator, if the northern potential $a_\mu^{(N)}(x)$ leaves the phase unchanged, then the southern potential $a_\mu^{(S)}(x)$ will have to rotate the phase by an amount equal to $4\pi e\tilde{e}$ on going once around the equator. We can thus imagine constructing the topology corresponding to a monopole, of unit charge $\tilde{e} = 1/(2e)$ say, as follows. We take a direct product of S^2 with S^1, i.e. Figure 2.8b', and cut it into two halves as depicted in Figure 2.8c'. We then make a twist on the southern half by rotating the attached circle once around while moving once around the equator, in the manner shown in Figure 2.8d', before rejoining the two pieces. The result is what in mathematics is called a nontrivial circle bundle with Chern class 1 over the sphere S^2 (see later in Section 3.4); and it is this topological structure which characterizes the magnetic monopole.

Let us compare the above construction with the familiar procedure for constructing the Möbius band. If we take a circular strip — i.e. the product of a circle S^1 with a line segment — cut it open, make a twist on one side, then stick back together, we obtain a Möbius band, as illustrated in Figure 2.8a-e; the steps are in one–one correspondence with those in Figure 2.8a'-e' for constructing the monopole bundle. Now, because of the twist, the Möbius band of Figure 2.8e, in contrast to the original circular strip of Figure 2.8b, cannot be shrunk to a point by any continuous deformation of the circle, as if there is something inside the circle which prevents it from doing so; hence

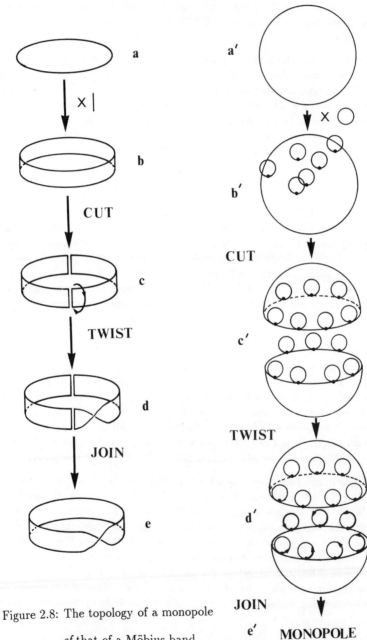

Figure 2.8: The topology of a monopole
cf that of a Möbius band.

the term *topological obstruction*. Similarly, again because of the twist, the nontrivial monopole bundle also cannot be shrunk to a point by any continuous deformation of the sphere, and the topological obstruction inside the sphere which prevents it from doing so is the magnetic monopole.

Having learnt that in the presence of a monopole the electromagnetic potential $a_\mu(x)$ has to be patched, one expects that the same will apply to the potential in a nonabelian theory. The generalization of the patching procedure itself to nonabelian theories is straightforward. Again, one introduces a separate gauge potential in each patch, only requiring that in the overlap region between any two patches (A) and (B) the gauge potentials are related by a patching transformation of the form (1.2.7) specified by a *patching function* S_{BA}:

$$A_\mu^{(B)}(x) = S_{BA}(x)A_\mu^{(A)}(x)S_{BA}^{-1}(x) - (i/g)\partial_\mu S_{BA}(x)\, S_{BA}^{-1}(x). \qquad (2.3.11)$$

However, since monopoles in nonabelian theories were defined above in terms of the phase factors $\Phi(C)$, we have first to clarify how these latter quantities behave under patching before we can specify the patching conditions for nonabelian monopoles.

2.4 Phase Transport with Patched Potentials

In the situation when the gauge potential is patched, one ought in principle to re-examine anew the meaning of all quantities and concepts which have been defined in terms of the potential. Local quantities such as the field tensor $F_{\mu\nu}(x)$ which depends on the gauge potential only at a point and its neighbourhood are not affected. Thus, given a point x in a given patch, (A) say, one can define $F_{\mu\nu}^{(A)}$ at x according to (1.2.12) in terms of the potential $A_\mu^{(A)}(x)$. For a point in the overlap region between two patches (A) and (B), one can define $F_{\mu\nu}^{(A)}$ and $F_{\mu\nu}^{(B)}$ respectively in terms of $A_\mu^{(A)}$ and $A_\mu^{(B)}$, and the two potentials being related by a patching gauge transformation, so will the corresponding field tensors. Thus, it follows from (1.2.13) that:

$$F_{\mu\nu}^{(B)}(x) = S_{BA}(x)\, F_{\mu\nu}^{(A)}(x)\, S_{BA}^{-1}(x). \qquad (2.4.1)$$

In particular, for the abelian theory, the two quantities $f_{\mu\nu}^{(A)}$ and $f_{\mu\nu}^{(B)}$ will have the same value in the overlap region, meaning of course that the field tensor need not even be patched in this case. This last remark will be of significance later in deriving the equations of motion of monopoles.

The place where one needs to be careful is in parallel phase transport over paths which traverse more than one patch. For a path which lies entirely

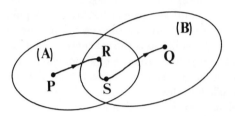

Figure 2.9: Illustration for phase transport with patched potentials (I).

within patch (A), such as PR in Figure 2.9, we can of course write the phase factor entirely in terms of the gauge potential $A_\mu^{(A)}$ in (A), thus

$$\Phi_{RP}(\Gamma) = P \exp ig \int_P^R A_\mu^{(A)}(x)\, dx^\mu. \qquad (2.4.2)$$

However, for a path such as PQ in Figure 2.9 which starts at P exclusively in (A) and ends at Q exclusively in (B), more care is needed. One can proceed as follows. One selects a point, say R, in the overlap region $(A) \cap (B)$ such that PR lies entirely in (A) and RQ lies entirely in (B). Then the phase factor Φ_{RP} can be written as in (2.4.2). Similarly, one can write the phase factor Φ_{QR} in terms of the potential $A_\mu^{(B)}$:

$$\Phi_{QR}(\Gamma) = P \exp ig \int_R^Q A_\mu^{(B)}(x)\, dx^\mu. \qquad (2.4.3)$$

We note, however, that Φ_{RP} represents parallel phase transport from P to R in the (A) gauge choice, while Φ_{RQ} that from R to Q but in the (B) gauge choice, and phases in the two gauge choices are related by the patching gauge transformation parametrized by $S_{BA}(R)$. Therefore, in effecting parallel phase transport from P to Q, one has to take account of this change of gauge at R and write the phase factor Φ_{QP} as:

$$\Phi_{QP}(\Gamma) = \Phi_{QR}(\Gamma)\, S_{BA}(R)\, \Phi_{RP}(\Gamma). \qquad (2.4.4)$$

Superficially, the expression seems to depend on the choice of the point R but it actually does not, as can be seen as follows. Suppose we choose another point S in $(A) \cap (B)$ and write:

$$\Phi'_{QP}(\Gamma) = \Phi_{QS}(\Gamma)\, S_{BA}(S)\, \Phi_{SP}(\Gamma), \qquad (2.4.5)$$

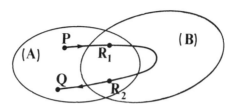

Figure 2.10: Illustration for phase transport with patched potentials (II).

with Φ_{SP} defined in terms of $A_\mu^{(A)}$ and Φ_{QS} defined in terms of $A_\mu^{(B)}$. This can also be written as:

$$\Phi'_{QP}(\Gamma) = \Phi_{QS}(\Gamma) S_{BA}(S) \left\{ P \exp ig \int_R^S A_\mu^{(A)}(x)\, dx^\mu \right\} \Phi_{RP}(\Gamma). \qquad (2.4.6)$$

One notes, however, that under the patching gauge transformation (2.3.11), one has according to (1.2.17):

$$P \exp ig \int_R^S A_\mu^{(A)}(x)\, dx^\mu = S_{BA}^{-1}(S) \left\{ P \exp ig \int_R^S A_\mu^{(B)}(x)\, dx^\mu \right\} S_{BA}(R),$$
$$(2.4.7)$$

which when substituted into (2.4.6) yields exactly again (2.4.4), meaning that the phase factor there is in fact independent of the choice of R in $(A) \cap (B)$.

We have assumed above that Γ is such that there exists a point R on Γ in $(A) \cap (B)$ such that PR lies entirely in (A) and RQ lies entirely in (B). There are of course other paths winding in and out of the overlap region for which this will not be true. One can, however, cut up any path into several pieces each lying entirely in one patch with its end-point in the overlap region. For example, the path in Figure 2.10 can be cut up into three pieces: PR_1, R_1R_2, and R_2Q, where PR_1 lies entirely in (A), R_1R_2 entirely in (B) and R_2Q entirely in (A) again, while R_1 and R_2 are both in the overlap region $(A) \cap (B)$. We can then write the phase factor $\Phi_{QP}(\Gamma)$ as:

$$\Phi_{QP}(\Gamma) = \Phi_{QR_2}(\Gamma) S_{AB}(R_2) \Phi_{R_2R_1}(\Gamma) S_{BA}(R_1) \Phi_{R_1P}(\Gamma), \qquad (2.4.8)$$

where Φ_{R_1P} is defined in terms of $A_\mu^{(A)}$, $\Phi_{R_2R_1}$ in terms of $A_\mu^{(B)}$ and Φ_{QR_2} in terms of $A_\mu^{(A)}$, and $S_{AB} = S_{BA}^{-1}$. Again it will be seen that Φ_{QP} so defined will not depend on R_1 or R_2 provided that they are both within the overlap region. From these examples, it is then clear that phase factors can similarly be defined for any path traversing the patches (A) and (B), or indeed any number of additional patches.

2.5 Potentials for Nonabelian Monopoles

Having defined parallel phase transport when the potential is patched, we can now examine what happens to the potential in the presence of a non-abelian monopole. Consider for illustration the simplest example of a theory with gauge group $SO(3)$, which contains already the essential features. We recall the manner in which an $SO(3)$-monopole was defined. If the procedure depicted in Figure 2.1 leads to a curve Γ which winds only once around the gauge group $SO(3)$, or equivalently only half-way round $SU(2)$ which is a double cover of $SO(3)$, we say then that the surface Σ in the figure encloses a monopole. In the anticipation that the presence of a monopole will require the potential to be patched, let us cover Σ again by the two patches (N) and (S) of (2.3.7) and introduce a gauge potential A_μ in each patch. Without loss of generality, let us choose the point P_0 of Figure 2.1 in the overlap region, say on the equator, which we identify as the loop C_t, say for $t = t_e$. North of the equator, we evaluate the phase factor $\Phi^{(N)}(C_t)$ in terms of the northern potential $A_\mu^{(N)}$, and south of the equator, $\Phi^{(S)}(C_t)$ in terms of the southern potential $A_\mu^{(S)}$. These phase factors, being by definition 2×2 unitary matrices, can be considered either as elements of $SU(2)$ or as elements of $SO(3)$, the difference being only that in the latter case matrices with opposite signs are to be regarded as the same group element. We choose here to regard the phase factors as elements of $SU(2)$.

Starting at $t = 0$ then, where $\Phi^{(N)}(C_0)$ is the unit matrix, the phase factor $\Phi^{(N)}(C_t)$ traces as t varies a continuous curve in the group $SU(2)$ until it reaches $t = t_e$. At $t = t_e$ one makes a patching transformation and transforms $\Phi^{(N)}(C_t)$ to $\Phi^{(S)}(C_t)$. From $t = t_e$ onwards, the phase factor $\Phi^{(S)}(C_t)$ again traces a continuous curve as t varies until t reaches 2π, where it becomes again the unit matrix and joins up with $\Phi^{(N)}(C_0)$. One sees thus that the only value of t at which the matrix function $\Phi(C_t)$ need not be continuous is at $t = t_e$. Hence, in order that the curve Γ traced out by $\Phi(C_t)$ for $t = 0 \to 2\pi$ should wind only half-way around $SU(2)$, we want:

$$P \exp ig \oint_{\text{equator}} A_\mu^{(N)}(x)\, dx^\mu = -P \exp ig \oint_{\text{equator}} A_\mu^{(S)}(x)\, dx^\mu, \qquad (2.5.1)$$

where the line integrals on both sides are of course taken in the same direction. This is then the patching condition on the gauge potential A_μ which typifies an

$SO(3)$ monopole with charge $-$ in analogy with the condition (2.3.10) for the monopole with charge $\tilde{e}$ in the abelian case. We notice that we have found it convenient to state the condition (2.5.1) not in the actual gauge group $SO(3)$ itself but in its covering group $SU(2)$. This is quite in analogy with the abelian condition (2.3.10) which was stated not in terms of the $U(1)$ element or the exponential phase factor, but in terms of the phase. The latter is an element of the real line $\mathbb{R}$, or, in other words, of the non-compact covering group of the compact gauge group $U(1)$, as explained in Section 1.4. As we shall see later, this tactic of utilizing the covering group of the actual gauge group will be found useful also in other similar situations.

As for the magnetic charge before, it is instructive at this point to write down an explicit example of a potential describing a nonabelian monopole. This is not difficult. The following, for n an odd integer, is already an example:

$$A_x^{(N)} = \frac{nt_3}{2g}\frac{y}{r(r+z)}, \quad A_y^{(N)} = -\frac{nt_3}{2g}\frac{x}{r(r+z)}, \quad A_z^{(N)} = A_t^{(N)} = 0, \qquad (2.5.2)$$

$$A_x^{(S)} = -\frac{nt_3}{2g}\frac{y}{r(r-z)}, \quad A_y^{(S)} = \frac{nt_3}{2g}\frac{x}{r(r-z)}, \quad A_z^{(S)} = A_t^{(S)} = 0, \qquad (2.5.3)$$

where we have simply replaced in the potential for the static magnetic monopole displayed in (2.3.8) and (2.3.9) the magnetic charge $\tilde{e}$ by $nt_3/(2g)$, with t_3 being $\frac{1}{2}$ times the third Pauli matrix:

$$t_3 = \frac{1}{2}\begin{pmatrix} 1 & 0 \\ 0 & -1 \end{pmatrix}. \qquad (2.5.4)$$

The presence of t_3 in the formulae above makes $A_\mu^{(N)}$ and $A_\mu^{(S)}$ elements of the $\mathfrak{su}(2)$ algebra, as the potential ought to be for an $SO(3)$ theory. However, the fact that the components all point in the same direction in internal symmetry space means they all commute and behave essentially as an abelian potential. The relation (2.3.10) therefore implies that:

$$\oint_{\text{equator}} \{A_\mu^{(N)} - A_\mu^{(S)}\}dx^\mu = 2n\pi t_3/g, \qquad (2.5.5)$$

or that the patching condition (2.5.1) is satisfied for n odd, as expected for the surface Σ to enclose a monopole.

The potentials of (2.5.2) and (2.5.3) for n even do not correspond to monopoles, since the phase factors Φ along the equator for (N) and (S) will then have the same sign and do not therefore satisfy the criterion (2.5.1) for an $SO(3)$ monopole. At first sight, this may seem surprising since the potential is still patched, whereas the absence of a monopole means topological equivalence to the vacuum and ought therefore to be representable by an unpatched potential. We must not forget, however, that the potential is only seen to be patched when all its components are chosen to point in the third internal symmetry direction. The gauge group being $SO(3)$, there are gauge transformations which rotate the potential in internal symmetry space, so that (2.5.2) and (2.5.3) for n even may in fact be gauge equivalent to an unpatched potential having no fixed internal symmetry direction.

It is instructive to construct an explicit example of such a gauge transformation. We shall do so for the potential (2.5.2) and (2.5.3) in the special case of $n = 2$, which should be sufficient to illustrate the essential features. We note first that in a patched system, one can choose to effect different gauge transformations in different patches. Let us then transform the gauge potentials in the northern and southern hemispheres respectively as follows:

$$A_\mu^{\prime(N)} = \xi A_\mu^{(N)}\xi^{-1} - (i/g)\partial_\mu\xi\,\xi^{-1}, \tag{2.5.6}$$

$$A_\mu^{\prime(S)} = \eta A_\mu^{(S)}\eta^{-1} - (i/g)\partial_\mu\eta\,\eta^{-1}, \tag{2.5.7}$$

with

$$\xi = \begin{pmatrix} \cos\frac{\theta}{2} & -\sin\frac{\theta}{2}e^{-i\phi} \\ \sin\frac{\theta}{2}e^{i\phi} & \cos\frac{\theta}{2} \end{pmatrix}, \tag{2.5.8}$$

$$\eta = \begin{pmatrix} \cos\frac{\theta}{2}e^{-i\phi} & -\sin\frac{\theta}{2} \\ \sin\frac{\theta}{2} & \cos\frac{\theta}{2}e^{i\phi} \end{pmatrix}, \tag{2.5.9}$$

where θ and ϕ are the usual polar coordinates. It is straightforward to check by direct calculation, separately in each patch, that one obtains as a result:

$$A_i^{\prime(N)} = A_i^{\prime(S)} = -\frac{1}{g}\epsilon_{ijk}\frac{x^k}{r^2}t_j, \quad A_t^{\prime(N)} = A_t^{\prime(S)} = 0. \tag{2.5.10}$$

The fact that in (2.5.10) the northern and southern potentials coincide shows already that in the new gauge, the potential is no longer patched. Alternatively, one can see this by gauge transforming the patching function. In the old gauge with the potential given by (2.5.2) and (2.5.3), the patching function S_{SN} relating $A_\mu^{(S)}$ to $A_\mu^{(N)}$ via (2.3.11) is easily seen to be:

$$S_{SN} = \exp int_3\phi = \begin{pmatrix} e^{in\phi/2} & 0 \\ 0 & e^{-in\phi/2} \end{pmatrix}. \tag{2.5.11}$$

Under the gauge transformations ξ and η in respectively the northern and southern hemispheres, the patching function transforms as:

$$S'_{SN} = \eta S_{SN}\xi^{-1}. \tag{2.5.12}$$

For $n = 2$ and ξ and η as exhibited in (2.5.8) and (2.5.9), this is easily seen to yield the identity matrix as the new patching function, meaning that indeed no patching is needed. The fact that the potential is unpatched in the new gauge shows also that it is topologically equivalent to the vacuum as anticipated, since there is nothing now to stop the potential from being continuously deformed to zero.

One notes, however, that the potential in (2.5.10) no longer points constantly in the third internal symmetry direction. Indeed, its direction in internal space not only depends on the space-time component labelled by i, but also depends on the space-time point x. The potential covers once all directions in internal symmetry space as the point x wanders once over the unit sphere in ordinary space, and is for this reason often described as a *hedgehog*. Projecting from this example, it is not hard to imagine that had one started with a potential in (2.5.2) and (2.5.3) with another even value of n, one could effect a similar transformation to remove the necessity for patching at the price of obtaining a potential in the new gauge which covers all directions in internal symmetry space $n/2$ times over as x moves once around the unit sphere in ordinary space. One sees then that the patched potential of (2.5.2) and (2.5.3) for even n are all indeed topologically equivalent to the vacuum. However, for n odd, no gauge transformation of this type can remove patching from the gauge potential because of the monopole. These observations will be useful in the next section when we consider symmetry breaking and the embedding of monopoles of a lower symmetry into a gauge theory with a higher symmetry.

2.6   Monopoles in Broken Symmetries

In practical applications of gauge theories, one often has to deal with so-called *spontaneously broken symmetries*. These are usually obtained by the introduction of Higgs scalar fields, ϕ say, with a potential $V(\phi)$ whose minima are not at $|\phi| = 0$, so that although the action remains invariant, the symmetry is 'broken' by the vacuum acquiring a preferred direction in internal symmetry space specified by a nonvanishing ϕ. In this section, we are interested in what happens to the monopoles of a gauge theory when the symmetry is spontaneously broken in this way.

Consider first the simple example of a $U(1)$ theory spontaneously broken by a potential $V(\phi)$ depending on a complex Higgs field ϕ, such that

$$V(\phi) = \text{min.}, \text{ for } |\phi| = \phi_0 > 0. \tag{2.6.1}$$

In order to minimize the total energy, we would require not only the potential to be minimum as in (2.6.1) but also the kinetic energy of ϕ to be minimum, which gives:

$$D_\mu\phi = (\partial_\mu - ieA_\mu)\,\phi = 0. \tag{2.6.2}$$

In the presence of a monopole, however, the condition (2.6.1) cannot be satisfied everywhere. To see this, we recall first that in any region surrounding a monopole the gauge potential has to be patched. Without loss of generality, let us specify the patches as in (2.3.7), and represent the Higgs field as $\phi^{(N)}$ and $\phi^{(S)}$ respectively in the northern and southern patches, such that along the equator:

$$\phi^{(S)} = S_{SN}\,\phi^{(N)}, \tag{2.6.3}$$

where $S_{SN}(x)$ is the patching function giving the transformation from the northern gauge to the southern gauge. We now see that on any sphere enclosing a monopole, the Higgs field must have at least one zero. Indeed, suppose that this is not the case, then $\hat\phi^{(N)} = \phi^{(N)}/|\phi^{(N)}|$ would exist everywhere in (N) and define a continuous map, say $\Gamma^{(N)}$, of the equator on to the unit circle S^1. By continuously deforming the equator into the north pole, the circuit $\Gamma^{(N)}$ contracts continuously to a point. In the same way, one sees that $\Gamma^{(S)}$ similarly defined can also be contracted continuously to a point. Now these two statements taken together contradict the original assertion that our sphere contains a monopole so that the function S_{SN} relating $\Gamma^{(N)}$ to $\Gamma^{(S)}$ has to belong to a nontrivial homotopy class on S^1. One has to conclude therefore that our assumption was wrong, or that ϕ must have a zero somewhere on the sphere surrounding the monopole. Given now that ϕ has at least one zero on every sphere surrounding a monopole, it follows that a monopole here must be attached to a string of such zeros along which — and by continuity for a finite distance in its proximity — the Higgs field must differ from its vacuum value $|\phi| = \phi_0$. This means this string of zeros of ϕ must carry energy, and have physical significance, in contrast to the Dirac string mentioned earlier, which was only a gauge artifact.

Next, consider the condition (2.6.2). This implies the relation:

$$[D_\mu, D_\nu]\phi = F_{\mu\nu}\phi = 0. \tag{2.6.4}$$

Hence to satisfy (2.6.2), $F_{\mu\nu}$ will have to vanish unless ϕ does. But $F_{\mu\nu}$ cannot vanish everywhere on a sphere surrounding a monopole since the total flux emerging from that sphere must add up to 4π times the magnetic charge enclosed inside. Clearly, the most energetically favourable arrangement would be to concentrate all the magnetic flux about the line where ϕ is zero.

We conclude therefore that in broken $U(1)$ symmetry, a monopole cannot exist isolated but must be attached to at least one massive string along which the Higgs field ϕ vanishes, which string is at the same time also a *vortex line* carrying a quantized magnetic flux. This vortex line can terminate only at another magnetic charge of the opposite sign, and since the energy carried by it will increase with its length, its tendency is to pull the charges together and annihilate them unless they are kept apart by other means.

Let us next turn to theories with gauge groups other than $U(1)$. Two situations can then occur with rather different consequences, depending on whether the symmetry is completely or only partially broken by the Higgs mechanism. Again, consider first the simplest example of a theory with gauge group $SO(3)$, which for convenience we shall here refer to as isospin. If we introduce one isotriplet of Higgs fields, ϕ^i say for $i = 1, 2, 3$, in the adjoint representation of the gauge group, with a potential whose minima are not at $\phi = 0$, then the vacuum will acquire a preferred direction in isospin space specified by a non-vanishing isovector ϕ_0 representing a particular minimal value of $V(\phi)$. The original $SO(3)$ gauge symmetry will then be broken since the vacuum is no longer invariant under all isospin rotations. However, the symmetry is only partially broken in that the system is still invariant under those isospin rotations which leave the vacuum invariant, namely the *little group* of $U(1)$ rotations about the axis specified by the isovector ϕ_0. On the other hand, if we introduce instead two triplets of Higgs fields, say ϕ^i and ψ^i, with a potential $V(\phi, \psi)$ whose minima occur at:

$$|\phi| > 0, |\psi| > 0, \text{ and } (\phi \cdot \psi) \neq 0, \qquad (2.6.5)$$

then the symmetry will be completely broken since once we have specified a particular vacuum satisfying (2.6.5), no isospin rotation will leave it invariant.

The situation for monopoles in a completely broken symmetry is similar to the earlier example of the broken $U(1)$ case. In analogy with the arguments above where we showed that on any sphere surrounding a $U(1)$ monopole, the Higgs field ϕ must have at least one zero, one can similarly demonstrate that on any sphere surrounding an $SO(3)$ monopole, two isovectors ϕ and ψ cannot remain everywhere linearly independent. That being the case, the Higgs fields in the above example must then depart from their vacuum values as specified by (2.6.5) at at least one point on any sphere surrounding the monopole. In other words, a monopole must again be attached to a physical energetic string which can be terminated only at another monopole of the 'opposite' charge (which, we recall, in the case of $SO(3)$, means the same charge $-$, since a monopole is here its own anti-particle). Further, in analogy with the abelian statement that the string should carry all the flux emerging from the monopole, here, since 'flux' according to the discussion in Section 1.3 is no longer unambiguously defined, we have instead the parallel statement that the phase factor $\Phi(C)$ defined as in (1.3.3) will depart appreciably from its trivial value of unity only for C passing through the vicinity of this string. In particular, for an infinitesimal loop around the string, $\Phi(C)$ will take on the value -1, so that we can again interpret the string as some sort of vortex line. The proof of these statements are very similar in spirit to that of their analogues in the abelian case, though a little more complicated, and will not therefore be repeated.

Instead, let us turn to the more interesting situation when the symmetry is only partially broken, as in the example considered above of an $SO(3)$ symmetry spontaneously broken by a single triplet of Higgs fields. In the presence of an $SO(3)$ monopole, the gauge potential will have to be patched, say again as in (2.3.7), in which case, we shall again have to introduce Higgs fields $\phi^{(N)}$ and $\phi^{(S)}$ separately for each patch to be related as in (2.6.3), except that the patching function S_{SN} is now a 3×3 orthogonal matrix. Equivalently, in the $su(2)$ notation adopted in previous sections for $SO(3)$, we may write instead:

$$(\phi^{(S)} \cdot \tau) = S_{SN} \, (\phi^{(N)} \cdot \tau) \, S_{SN}^{-1}, \qquad (2.6.6)$$

with S_{SN} a 2×2 unitary matrix. Though formally similar to the abelian case, the equation (2.6.3) or (2.6.6) is in fact very different in content. Indeed, if we try now to repeat our previous argument in the abelian theory deducing that ϕ must deviate from its vacuum value somewhere on any sphere surrounding the monopole, we shall fail to do so because $\Gamma^{(N)}$ and $\Gamma^{(S)}$ defined as before will now be maps of the equator into the unit sphere, which are always contractible to points in any case, and thus in no way contradictory to the requirement that the patching function S_{SN} should belong to a nontrivial homotopy class on the gauge group. There is in fact no difficulty in constructing explicitly Higgs fields satisfying (2.6.1) everywhere around a monopole. We need only take:

$$\phi^{(N)} = \phi^{(S)} = \phi_0 \begin{pmatrix} 0 \\ 0 \\ 1 \end{pmatrix} \qquad (2.6.7)$$

and the patching function (2.5.11). Being just a rotation about the third isospin axis, this S_{SN} will leave the Higgs field above invariant. However, so long as n in (2.5.11) is odd, S_{SN} is a valid patching function for an $SO(3)$ monopole, as seen already in the example of the static potential in (2.5.2) and (2.5.3). The conclusion is thus that in contrast to the case of a completely broken symmetry, an $SO(3)$ monopole in a partially broken theory need not be attached to an energetic string but can exist isolated since the Higgs field need nowhere depart from its vacuum value. We notice further that the potential (2.5.2) and (2.5.3) satisfy also the condition:

$$D_\mu \phi = \partial_\mu \phi - ig[A_\mu, (\phi \cdot \tau/2)] = 0, \qquad (2.6.8)$$

for minimizing the kinetic energy. There is thus also no concentration of 'flux' in one particular direction since the phase factor (1.3.3) evaluated with (2.5.2) and (2.5.3) can deviate from unity without a loop passing through the vicinity of some 'vortex line', in contrast to the case above for completely broken symmetry.

One may feel that the above example is very special in that both the Higgs field and all components of the gauge potential point in the same isospin direction at all spacetime points. This is in fact not the case, for given any Higgs field pointing in different directions at different spacetime points, one can always effect a (patched) gauge transformation to rotate it to a fixed isospin direction, say, without loss of generality, the third direction as above. Then in order to satisfy the condition (2.6.8) for minimizing the kinetic energy, it is sufficient to have all components of the potential A_μ pointing in the same isospin direction. Choosing then to work in this particularly simple gauge, it becomes transparent what actually happens to an $SO(3)$ monopole when the symmetry is partially broken. Since all fields now point in the third isospin direction, the patching function can represent only a rotation about the third isospin axis, and being by definition single-valued, it has to be of the form (2.5.11) with n an integer. The patching condition (2.5.1) for an $SO(3)$ monopole then implies that n must be odd. This means that it becomes just the patching function for an abelian monopole of odd charge n which is indeed the case in the specific example of the static potential (2.5.2) and (2.5.3). In other words, one sees that when the symmetry is thus broken, the $SO(3)$ monopole *reduces* to a $U(1)$ monopole of odd abelian (magnetic) charge n. This last statement is gauge invariant, and although it is easiest to see in the particular gauge when the Higgs field and the potential are both pointing in the same isospin direction, it need not always be so represented. In the preceding section, we have already seen an explicit example in (2.5.6) and (2.5.7) of a gauge transformation which turns the above representation of such a reduced monopole of the residual $U(1)$ into one in which the Higgs field and gauge potential point in varying isospin directions, but the physical content remains of course the same.

It is interesting to remark here also about the monopoles with even charges n of the residual symmetry. We have seen already in the section above that the potential for these can be gauge transformed through an $SO(3)$ transformation into the hedgehog form (2.5.10) which is unpatched. In our present language, this means that an $SO(3)$ potential with monopole charge $+$, namely that of the vacuum, will under symmetry breaking, reduce to even-charged monopoles of the residual $U(1)$ symmetry. That this can be the case is not really surprising, since the gauge group $SO(3)$ is larger than the residual symmetry $U(1)$. There are thus topological structures which can be unwound by $SO(3)$

transformations, thus becoming trivial, but cannot be similarly unwound by the more restricted $U(1)$ group of transformations. Indeed, for those readers who have come across the 't Hooft–Polyakov work, this is exactly the manner in which these authors obtained nonsingular soliton solutions in the partially broken $SO(3)$ theory which are at the same time monopoles of the residual $U(1)$ symmetry. They are $U(1)$ monopoles embedded in the vacuum sector of the bigger nonabelian $SO(3)$ symmetry, and *not* nonabelian $SO(3)$ monopoles in our present language. They ought not to be confused with the latter.

Although we have worked explicitly with only the simplest examples, the arguments can without great difficulty be extended. In general, it can be shown that a monopole of a spontaneously broken gauge symmetry with gauge group G can exist isolated (i.e. without being attached to an energetic string) if and only if in that homotopy class of closed circuits of G associated with that monopole charge, there are circuits which lie entirely in the residual subgroup H. In that case, this G monopole reduces to an H monopole. (See also Section 3.4.)

In spite of the knowledge of the above general result, it may still be worthwhile giving the explicit construction for the standard electroweak theory as a practical example, which has in the past caused some confusion. As explained before in Section 1.4, the gauge group in the present electroweak theory is $U(2)$ and not $SU(2) \times U(1)$, but we shall consider both these theories simultaneously. Since $SU(2) \times U(1)$ is a double cover of $U(2)$, close circuits in $U(2)$ belonging to the homotopy class $2m$ are also closed circuits in $SU(2) \times U(1)$, so that homotopy classes of the latter group may be labelled by even integers. Given then a monopole with $m \neq 0$, we ask whether it needs to be attached to a vortex line when the symmetry is broken by a potential $V(\phi)$ satisfying:

$$V(\phi) = \text{min.}, \text{ for } |\phi|^2 = \phi^\dagger \phi = \phi_0^2 \neq 0, \qquad (2.6.9)$$

where ϕ is an $SU(2)$ doublet.

As before, we first examine whether on a sphere surrounding the monopole the Higgs field ϕ can remain always at the minimum of the potential V. Let

$$\phi^{(N)} = \phi^{(S)} = \phi_0 \begin{pmatrix} 0 \\ 1 \end{pmatrix} \qquad (2.6.10)$$

in a gauge with the usual patching by two hemispheres, and a patching function for the case of $U(2)$:

$$S_{SN} = \exp -im\phi(\tau_3/2 + I/2) = \begin{pmatrix} e^{-im\phi} & 0 \\ 0 & 1 \end{pmatrix} \qquad (2.6.11)$$

or for the case of $SU(2) \times U(1)$:

$$S_{SN} = (\exp -im\phi\tau_3/2, \exp -im\phi I/2)$$
$$= \left(\begin{pmatrix} e^{-im\phi/2} & 0 \\ 0 & e^{im\phi/2} \end{pmatrix}, \begin{pmatrix} e^{-im\phi/2} & 0 \\ 0 & e^{-im\phi/2} \end{pmatrix} \right) \quad (2.6.12)$$

which leaves (2.6.10) invariant. In the notation of Figure 2.7, S_{SN} in (2.6.11) or (2.6.12) lies along the diagonal, and as $\phi = 0 \to 2\pi$, it traces a circuit winding m times round $U(2)$ or $m/2$ times around $SU(2) \times U(1)$. Our construction above represents therefore already a valid Higgs field on a sphere enclosing the monopole with charge m. This Higgs field (2.6.10) lies everywhere on the minimum of V. There is thus no need for an energetic string, and the monopole can exist isolated. As for the monopole 'flux', one sees that the minimization of the kinetic energy requires that

$$D_\mu \phi = i\Gamma_\mu \phi = 0, \quad (2.6.13)$$

where

$$\Gamma_\mu = g A_\mu^i(\tau_i/2) + g' B_\mu(-I/2), \quad (2.6.14)$$

which means that only that component of the potential lying in the little group of ϕ, namely:

$$a_\mu = -\frac{e}{g} A_\mu^3 + \frac{e}{g'} B_\mu, \quad (2.6.15)$$

with

$$e = gg'/(g^2 + g'^2)^{1/2}, \quad (2.6.16)$$

can be non-zero. To see whether 'flux' can leak out via this component alone, we take:

$$A_\mu^1 = A_\mu^2 = 0, \quad (2.6.17)$$

$$Z_\mu = \frac{e}{g'} A_\mu^3 + \frac{e}{g} B_\mu = 0, \quad (2.6.18)$$

and a_μ as given in (2.3.8) and (2.3.9) with $\tilde{e}$ replaced by $m/(2e)$. It can readily be checked that this gauge potential satisfies the patching condition:

$$\Gamma_\mu^{(S)} = S_{SN} \Gamma_\mu^{(N)} S_{SN}^{-1} - i S_{SN} \partial_\mu S_{SN}^{-1}. \quad (2.6.19)$$

The phase factor $\Phi(C)$ in (1.3.3) is now to be evaluated with Γ_μ in place of A_μ. It can readily be shown then by direct calculation that $\Phi(C)$ can deviate from unity for any loops, not necessarily those passing through the vicinity of some vortex line, meaning thus that the 'flux' can indeed leak out in any direction. Again, although the statements here are particularly simple to deduce in the gauge where all fields point in a fixed isospin direction, they remain valid in other gauges. Indeed, the reduction from $U(2)$ to $U(1)$ monopoles was historically first derived in the Wu-Yang hedgehog gauge similar to (2.5.10) in which the fields point in varying isospin directions and the above conclusion was therefore less transparent.

Chapter 3

Mathematical Interlude

So far we have deliberately avoided using mathematical concepts which may be unfamiliar to most physicists, preferring to derive all the results from physical first principles. The time has come, however, to introduce mathematics for the triple purpose of conciseness, clarity and guidance for future generalizations. We shall not repeat here in detail the sort of mathematics that most physicists know nowadays, but shall discuss only those parts that are not normally taught in physics graduate schools. Our aim is twofold. First, we wish to introduce such mathematical concepts as are necessary for a clearer understanding of the physical topics dealt with in this book. We shall try to do so in as simple a manner as we can manage, and to illustrate these concepts with equally simple examples familiar to physicists. Secondly, we wish to collect together for easy reference some related mathematical results in anticipation of possible future applications within the context of our subject. This material is scattered over a wide area in the mathematical literature and may be tedious to locate for those unfamiliar with it. For this reason, we think it is helpful to have at hand a concise summary of possibly useful known results, although it is obviously not easy to know exactly what is likely to be needed. This summary will be given mostly in the form of tables. We suggest that the reader, at first reading, go quickly through this chapter, omitting most tabulated details and those abstract points he (or she) may find difficult to digest, yet trying to familiarize himself with the contents so that he can refer back to them when the need arises. Although it may not be necessary to master every mathematical concept introduced here in order to attack the physical problems to be treated later, the reader, we believe, will find his understanding of the subject much enhanced when it is placed in a wider mathematical context.

We shall endeavour to be scrupulously correct in our statements, though we shall keep away from the formal presentation normally encountered in mathematics textbooks so difficult for physicists to follow. In the same spirit, no

Symbol	Meaning
$\exists$	there exists
$\forall$	for all
$\emptyset$	the empty set
$\mathbb{Z}$	the integers
$\mathbb{Z}_n$	the integers modulo n
$\mathbb{Q}$	the rational numbers
$\mathbb{R}$	the real numbers
$\mathbb{C}$	the complex numbers
$\Rightarrow$	implies
$\Leftrightarrow$	iff (if and only if)
$x \in X$	x is an element of X
$\square$	denotes the end of a proof

Table 3.1: Some current symbols

attempt whatsoever will be made to furnish formal proofs. Nor is there any pretence to completeness.

General elementary ideas about the following are assumed: general topology, manifolds, Lie algebras, Lie groups. We have included some standard textbooks in the bibliography at the end, as well as some we have found particularly useful.

From time to time we find it convenient to adopt some symbols in current mathematical usage, summarized in Table 3.1. Furthermore, the notation for functions is as follows:

$$f : \quad X \quad \to \quad Y$$
$$x \quad \mapsto \quad f(x)$$

denoting a map f from the space X to the space Y taking the point x to its image $f(x)$.

We also find it convenient to denote the range of a parameter as follows: $\theta = 0 \to 2\pi$, instead of the more usual $0 \leq \theta \leq 2\pi$.

Unless stated otherwise, we shall use the term *differentiable* to mean infinitely differentiable, but of course not necessarily analytic. The term *space* is generally taken to mean topological space; however, since we are interested almost solely in manifolds, we shall assume for our spaces all the 'good' properties of manifolds, except where the contrary is explicitly stated. By this slight abuse of language we shall be able to avoid having to stipulate repeatedly, e.g. locally connected, locally path connected, or locally compact spaces, without bringing in a differentiable structure unnecessarily.

3.1 Homotopy Groups

The most elementary and obvious property of a topological space X is the number of connected parts or components it has. The next such property, in a certain sense, is the number of holes X has. We have already seen in the preceding chapters how these properties of space enter in consideration of physical problems such as the outcome of the Aharonov–Bohm experiment and the possible values of the monopole charge. Mathematically, the importance of these properties is that they are topological invariants in the following sense: if X' is another topological space such that there exists a continuous one-to-one map $\phi : X \to X'$, with an inverse ϕ^{-1} which is also continuous, then X and X' have the same number of components and holes. (Such a ϕ is called a *homeomorphism*.) A descriptive way of saying this is that they are invariant under continuous deformations. Consider for example the conic sections: a hyperbola has two components and an ellipse has one hole. These two properties remain unchanged under suitable small changes of the parameters. What makes these topological invariants even more interesting is that higher-dimensional analogues exist which are also topological invariants.

First, let us define the zeroth homotopy set of X, denoted $\pi_0(X)$, as the set of path connected components[1] of X. Obviously, a *connected space* is one where π_0 is trivial, which we usually write as $\pi_0 = 0$.

For what follows, the most convenient way to define the higher homotopy groups $\pi_n(X)$, $n \geq 1$, is through loop space. For technical reasons, it is customary to fix a base point x_0 in X. Consider all the continuous closed loops in X beginning and ending at x_0, i.e. all continuous functions of the unit circle parametrized by $\theta = 0 \to 2\pi$ into X mapping 0 and 2π to x_0. The set of such based loops in X is called the *(pointed) (parametrised) loop space* of X, denoted ΩX. As a function space it can be made into a topological space in the usual way by giving it the compact open topology.

The set $\pi_1(X)$ is defined to be $\pi_0(\Omega X)$, the base point of ΩX being the constant loop at x_0, i.e. the map which takes all points of the unit circle to the base point x_0. The elements of $\pi_1(X)$ are therefore classes of closed loops in X which cannot be continuously deformed into one another. The composition of loops induces a group structure on $\pi_1(X)$, with identity being the constant loop at x_0. We can therefore call $\pi_1(X)$ the *first homotopy group* of X; it is also known as the *fundamental group* of X. For example, in the case where X is the gauge group G of a Yang–Mills theory, the elements of $\pi_1(G)$ are exactly the homotopy classes of closed curves Γ which were used in Sections 2.1 and

---

[1] The concepts of connectedness and path connectedness are identical in all cases of interest to us, e.g. when X is a manifold.

2.2 to define monopole charges. A space with $\pi_1 = 0$, namely having only one homotopy class of closed loops which can all be continuously deformed into one another and hence also to a point, is said to be *simply-connected*.

We shall take the higher homotopy groups as being defined recursively: $\pi_n(X) = \pi_{n-1}(\Omega X)$, $n \geq 1$. Notice that for $n \geq 1$, $\pi_n(X)$ is a group, the group action coming from the joining of two loops together to form a new loop, just as in the case of the fundamental group above. $\pi_0(X)$ in general is not a group. However, when X is a Lie group, then $\pi_0(X)$ inherits a group structure from X, because it can be identified with the quotient group of X by its identity-connected component. For example, the two components of $O(3)$ can be identified with the two elements of the group $\mathbb{Z}_2$, the component where the determinant equals 1 corresponding to 0 in $\mathbb{Z}_2$ and the component where the determinant equals -1 corresponding to 1 in $\mathbb{Z}_2$. Furthermore, for $n \geq 2$, $\pi_n(X)$ is always abelian. $\pi_1(X)$ is in general not abelian. For example, if X is a Riemann surface with genus > 1, then $\pi_1(X)$ is not abelian. However, the fundamental group of a Lie group is always abelian. For these results and many of those that follow, the reader will find a particularly interesting chapter in Steenrod (1951).

We note in passing that homotopy groups are in general difficult to calculate; quite often the only way is to do it case by case, and many still are unknown (e.g. the higher homotopy groups of S^2). The only really easy result is for product spaces:

$$\pi_n(X \times Y) = \pi_n(X) + \pi_n(Y), \quad n \geq 1 \qquad (3.1.1)$$

where on the right the product group is usually written additively because $\pi_n(X)$ is abelian when either X is a Lie group or $n \geq 2$. Another important general result on the homotopy sequence of fibre bundles will be mentioned after the latter are introduced (Section 3.4).

For topics dealt with in this book, we shall be interested mainly in the low order homotopy groups of spheres S^n and of classical and exceptional groups. We defer the discussions of the homotopy groups of the classical and exceptional Lie groups till the next section. For the spheres, we have the following results:

$$
\begin{aligned}
\pi_i(S^n) &= \begin{cases} 0 & \text{if } i < n \\ \mathbb{Z} & \text{if } i = n \end{cases} \\
\pi_i(S^1) &= 0 \quad\;\; \text{if } i > 1 \\
\pi_{n+1}(S^n) &= \mathbb{Z}_2 \quad \text{if } n \geq 3 \\
\pi_{n+2}(S^n) &= \mathbb{Z}_2 \quad \text{if } n \geq 2 \\
\pi_6(S^3) &= \mathbb{Z}_{12}
\end{aligned}
\qquad (3.1.2)
$$

From the theory of sphere bundles (see Section 3.4), we can deduce:

$$\pi_i(S^2) = \pi_{i-1}(S^1) + \pi_i(S^3) \quad \text{if } i \geq 2$$
$$\pi_i(S^4) = \pi_{i-1}(S^3) + \pi_i(S^7) \quad \text{if } i \geq 2$$
$$\pi_i(S^8) = \pi_{i-1}(S^7) + \pi_i(S^{15}) \quad \text{if } i \geq 2,$$

and the first of these relations give the following more succinct result:

$$\pi_i(S^3) = \pi_i(S^2) \quad \text{if } i \geq 3.$$

A result of Serre says that all the homotopy groups of spheres are in fact finite except $\pi_n(S^n)$ and $\pi_{4n-1}(S^{2n})$, $n \geq 1$.

Before we leave the homotopy groups we have to take a closer look at the fundamental group π_1 and its relation to covering spaces. Given a connected space X, a map $\pi : B \to X$ is called a *covering* if (i) $\pi(B) = X$, and (ii) for each $x \in X$, $\exists$ an open connected neighbourhood V of x such that each component of $\pi^{-1}(V)$ is open in B. The space B is called a *covering space*. Thus, for example, in cases we have already seen, the real line $\mathbb{R}$ is a covering of the group $U(1)$ and, as we shall further clarify in the next section, the group $SU(2)$ is a double cover of the group $SO(3)$. By considering closed loops in X and their coverings in B it is easily seen that the fundamental group $\pi_1(X)$ acts on the coverings of X. See also Section 3.4 and especially Example 1 there. If we further assume that the action is transitive, then we have the following nice result: *coverings of X are in 1–1 correspondence with normal subgroups of $\pi_1(X)$.*

Given a connected space X, there always exists a unique connected simply connected covering space $\widetilde{X}$, called the *universal covering space*. Furthermore, $\widetilde{X}$ covers all the other covering spaces of X. For the higher homotopy groups, one has

$$\pi_n(X) = \pi_n(\widetilde{X}), \quad n \geq 2. \tag{3.1.3}$$

3.2 Lie Groups and Lie Algebras

We assume the basic facts about Lie groups and Lie algebras are known. In particular, we shall not touch upon their representation theory. Here we merely point out certain useful facts which might be unfamiliar to physicists. Since we are dealing almost exclusively with Lie algebras and Lie groups, we shall occasionally omit the name 'Lie'.

Briefly a Lie group is both a group and an analytic manifold in which the group operation is analytic. We shall denote Lie groups here by upper case Roman letters. To each Lie group, say G, corresponds a Lie algebra, denoted here by the corresponding letter in lower case German Fraktur, say $\mathfrak{g}$, which may be defined in several equivalent ways. One convenient way is to identify $\mathfrak{g}$, as a vector space, with the tangent space $T_e(G)$ to G (as a manifold) at the identity e. Let X, $Y \in T_e(G)$, thought of as tangent vectors at e to two curves $x(t)$, $y(t)$ passing through e at $t = 0$. Then $\lim_{t \to 0}(1/t^2)x(t)y(t)x(t)^{-1}y(t)^{-1}$ is again a tangent vector at e, called the *commutator, Lie bracket* or *Lie product* $[X, Y]$. This product is bilinear, skew-symmetric and satisfies the Jacobi identity, making $T_e(G) = \mathfrak{g}$ a Lie algebra. The Jacobi identity in $\mathfrak{g}$ reflects the associative law in G. When we are dealing with matrix groups then the Lie algebras can be realized as matrix algebras, in which case the bracket is concretely given as the commutator of the matrices. In an abstract way this can always be done, i.e. the Lie algebra can be embedded in an associative algebra, called its *universal enveloping algebra*, in such a way that the Lie bracket is given as the commutator in the universal enveloping algebra.

Recall that a Lie algebra is also a vector space. Hence it possesses a basis. Elements of the basis of $\mathfrak{g}$ are called *(infinitesimal) generators* of the algebra as well as the generators of any group for which $\mathfrak{g}$ is the Lie algebra.

For every Lie group the above definition gives uniquely a Lie algebra. Given a Lie algebra, however, there may be more than one Lie group for which it is the Lie algebra[2]. These groups are then said to be *locally isomorphic*. For example, the 1-dimensional abelian Lie algebra $\mathbb{R}$ is the Lie algebra of the following four Lie groups (among many others): the additive group $\mathbb{R}$, $U(1) \cong \{z \in \mathbb{C} : |z| = 1\}$, the positive reals $\mathbb{R}^+$, and the non-zero reals $\mathbb{R}^\times$ (called the multiplicative group of $\mathbb{R}$, with 2 connected components), the last three with multiplication as group operation. In general, the connected groups are covering spaces for the 'smallest' one, i.e. the one with the largest fundamental group. Thus in the above example, $U(1)$ has $\mathbb{Z}$ as its fundamental group, and its universal cover is $\mathbb{R}$. In fact all locally isomorphic groups can be obtained by factoring the simply connected covering group by various discrete normal subgroups of its fundamental group.

As another elementary example of different groups corresponding to the same algebra, let us study in detail one that we have already met in Chapter 2 to which we shall often return. The trace-free skew hermitian 2×2 matrices constitute the Lie algebra $\mathfrak{su}(2)$. It has generators:

$$X_i = -\tfrac{1}{2}i\sigma_i,$$

---

[2]In infinite dimensions the group may not exist. For example, it is known, but the proof is difficult, that no Lie group corresponds to $\mathrm{Vect}_{\mathbb{C}}S^1$, the complexification (see later) of the Virasoro algebra (without central extension).

satisfying the commutation relation:

$$[X_i, X_j] = \epsilon_{ijk} X_k, \tag{3.2.1}$$

where σ_i are the standard Pauli matrices:

$$\sigma_1 = \begin{pmatrix} 0 & 1 \\ 1 & 0 \end{pmatrix}, \quad \sigma_2 = \begin{pmatrix} 0 & -i \\ i & 0 \end{pmatrix}, \quad \sigma_3 = \begin{pmatrix} 1 & 0 \\ 0 & -1 \end{pmatrix}. \tag{3.2.2}$$

Consider next the Lie algebra $\mathfrak{so}(3)$ of 3×3 skew-symmetric matrices, for which we may choose as basis:

$$Y_1 = \begin{pmatrix} 0 & 0 & 0 \\ 0 & 0 & 1 \\ 0 & -1 & 0 \end{pmatrix}, \quad Y_2 = \begin{pmatrix} 0 & 0 & 1 \\ 0 & 0 & 0 \\ -1 & 0 & 0 \end{pmatrix}, \quad Y_3 = \begin{pmatrix} 0 & 1 & 0 \\ -1 & 0 & 0 \\ 0 & 0 & 0 \end{pmatrix}.$$

Then we have

$$[Y_i, Y_j] = \epsilon_{ijk} Y_k,$$

showing that $\mathfrak{su}(2) \cong \mathfrak{so}(3)$.

However, the groups $SU(2)$ and $SO(3)$ are not isomorphic. Instead, there is a 2–1 map $SU(2) \to SO(3)$ in such a way that $SU(2)$ is a double cover of $SO(3)$. Explicitly, $SU(2)$ can be parametrized as follows:

$$U(\phi, \alpha, \beta) = \begin{pmatrix} \cos\phi \, \exp i\alpha & \sin\phi \, \exp -i\beta \\ -\sin\phi \, \exp i\beta & \cos\phi \, \exp -i\alpha \end{pmatrix}, \quad \begin{array}{l} 0 \le \phi < \tfrac{1}{2}\pi, \\ 0 \le \alpha, \beta < 2\pi \end{array},$$

and $SO(3)$ as follows in terms of the Euler angles:

$$O(\phi_1, \theta, \phi_2) = \begin{pmatrix} \cos\phi_1 & -\sin\phi_1 & 0 \\ \sin\phi_1 & \cos\phi_1 & 0 \\ 0 & 0 & 1 \end{pmatrix} \begin{pmatrix} 1 & 0 & 0 \\ 0 & \cos\theta & -\sin\theta \\ 0 & \sin\theta & \cos\theta \end{pmatrix} \begin{pmatrix} \cos\phi_2 & -\sin\phi_2 & 0 \\ \sin\phi_2 & \cos\phi_2 & 0 \\ 0 & 0 & 1 \end{pmatrix}$$

with $0 \le \phi_1 < 2\pi$, $0 \le \phi_2 < 2\pi$, $0 \le \theta < \pi$. Then the 2–1 map is

$$\pm U(\tfrac{1}{2}\theta, \tfrac{1}{2}(\phi_1 + \phi_2), -\tfrac{1}{2}(\pi + \phi_1 - \phi_2)) \mapsto O(\phi_1, \theta, \phi_2). \tag{3.2.3}$$

Thus, one sees that a gauge theory with gauge algebra $\mathfrak{su}(2)$ can have gauge group either $SO(3)$ or $SU(2)$, and the choice, as explained in Sections 1.4 and 2.2 above, depends on the particle and field content of the theory. We have seen also that the study of monopoles in the theory depends crucially on a proper identification of the relevant group.

From a group G, one obtains the Lie algebra by taking the tangent space to the group at its identity as explained above. Conversely, the exponential map goes from the Lie algebra $\mathfrak{g}$ back to the Lie group G. The exponential map of elementary analysis:

$$\exp:\ \mathbb{R} \to\ U(1)$$
$$x \mapsto\ \exp ix$$

is an example. For a general G, the subgroups generated by a vector X in $\mathfrak{g}$, of the form $\exp tX$, $t \in \mathbb{R}$, are called *one-parameter subgroups*. They are in the image of the exponential map. In finite dimensions, the exponential map is a homeomorphism from a neighbourhood of zero in $\mathfrak{g}$ to a neighbourhood of the identity in G. Outside of these neighbourhoods, however, it is not guaranteed to be either *injective* (i.e. mapping distinct elements to distinct images) or *surjective* (i.e. having image points covering the whole of G). The non-injectivity is obvious, even in the elementary example above where algebra elements x differing by integral multiples of 2π are mapped to the same element $\exp ix$ of the group. A simple example of the exponential map failing to be surjective is the group $SL(2,\mathbb{R})$ of real 2×2 matrices with unit determinant the algebra for which is given in the list below. It is easy to show that a matrix in the image of the exponential map must have a trace ≥ -2, and therefore this image cannot be the whole of $SL(2,\mathbb{R})$ which contains matrices with arbitrary traces. In infinite dimensions the situation can be much worse. For example, for $\mathrm{Diff}\, S^1$, the group of diffeomorphisms of the circle, it can be shown that the exponential map is neither injective nor surjective for any neighbourhood of the identity. A proof of this surprising result can be found in Pressley and Segal (1986).

Given a Lie algebra $\mathfrak{g}$ and an element $V \in \mathfrak{g}$, we have a linear transformation $U \mapsto [U, V]$ which is denoted $\mathrm{ad}\, V$. From this we can define a bilinear form on $\mathfrak{g} \times \mathfrak{g}$, called the *Killing form*, by

$$B(V_1, V_2) = \mathrm{tr}\,(\mathrm{ad} V_1\, \mathrm{ad} V_2)$$

in the sense that given any other two vectors $U_1,\ U_2 \in \mathfrak{g}$,

$$B(V_1, V_2)\,(U_1, U_2) = \mathrm{tr}\,([U_1, V_1]\,[U_2, V_2]),$$

which is a number. We say that $B(V_1, V_2)$ is non-degenerate if it vanishes only when either $V_1 = 0$ or $V_2 = 0$. A Lie algebra $\mathfrak{g}$ is said to be *semi-simple* if its Killing form is non-degenerate. $\mathfrak{g}$ is *simple* if it is nonabelian[3] and has no ideals except $\{0\}$ and $\mathfrak{g}$. A Lie group is *semi-simple* (*simple*) if its algebra is

---

[3]Some authors prefer to include the abelian case, i.e. say 'non-trivial' instead of 'non-abelian', but that would make say $U(1)$ simple but not semi-simple. The two definitions coincide when dimension > 1.

semi-simple (simple). An equivalent characterization for a semi-simple group is that it is locally isomorphic to a direct product of simple groups. So in a sense simple groups (and algebras) are the fundamental building blocks. It is well-known that Cartan gave in his thesis (1894) a complete classification of simple Lie algebras over $\mathbb{C}$. Instead of giving any details, we shall summarize the results in tables.

First some notation. In what follows, left superscript t denotes the transpose and a bar the complex conjugate of a matrix.

$GL(n, \mathbb{C}), GL(n, \mathbb{R})$: the group of complex, resp. real, $n \times n$ matrices with det $\neq 0$.

$SL(n, \mathbb{C}), SL(n, \mathbb{R})$: the group of complex, resp. real, $n \times n$ matrices with det $= 1$.

$U(p, q)$: the group of matrices g in $GL(p + q, \mathbb{C})$ leaving invariant the hermitian form
$$-z_1 \bar{z}_1 - \cdots - z_p \bar{z}_p + z_{p+1} \bar{z}_{p+1} + \cdots + z_{p+q} \bar{z}_{p+q}, \text{ i.e. } {}^t g I_{p,q} \bar{g} = I_{p,q}.$$

$U(n) = U(0, n) = U(n, 0)$: the group of $n \times n$ unitary matrices.

$SU(p, q)$: The group of matrices in $U(p, q)$ with det $= 1$.

$SU(n) = SU(0, n) = SU(n, 0)$: the group of $n \times n$ unitary matrices with det $= 1$.

$SO(n, \mathbb{C})$: the group of matrices g in $SL(n, \mathbb{C})$ leaving invariant the quadratic form
$$z_1^2 + \cdots + z_n^2, \text{ i.e. } {}^t g g = I_n.$$

$SO(p, q)$: the group of matrices g in $SL(p+q, \mathbb{R})$ leaving invariant the quadratic form
$$-x_1^2 - \cdots - x_p^2 + x_{p+1}^2 + \cdots + x_{p+q}^2, \text{ i.e. } {}^t g I_{p,q} g = I_{p,q}.$$

$SO(n) = SO(0, n) = SO(n, 0)$: the group of $n \times n$ orthogonal matrices.

$Sp(n, \mathbb{C}), Sp(n, \mathbb{R})$: the group of matrices g in $GL(2n, \mathbb{C})$, resp. $GL(n, \mathbb{R})$, such that ${}^t g J_n g = J_n$.

$Sp(p, q)$: the group of matrices g in $Sp(p + q, \mathbb{C})$ such that ${}^t g K_{p,q} \bar{g} = K_{p,q}$.

$Sp(n) = Sp(0, n) = Sp(n, 0)$: the group of $n \times n$ symplectic matrices.

Here we have used:

$$I_{p,q} = \begin{pmatrix} -I_p & 0 \\ 0 & I_q \end{pmatrix}, \quad J_n = \begin{pmatrix} 0 & I_n \\ -I_n & 0 \end{pmatrix}, \quad K_{p,q} = \begin{pmatrix} I_{p,q} & 0 \\ 0 & I_{p,q} \end{pmatrix},$$

and I_n =the identity $n \times n$ matrix.

The corresponding classical algebras are the following.

$\mathfrak{gl}(n, \mathbb{C}), \mathfrak{gl}(n, \mathbb{R})$ ={all $n \times n$ complex, resp. real, matrices}

$\mathfrak{sl}(n, \mathbb{C}), \mathfrak{sl}(n, \mathbb{R})$ ={all $n \times n$ trace-free complex, resp. real, matrices}

$$\mathfrak{u}(p, q) = \left\{ \begin{pmatrix} Z_1 & Z_2 \\ {}^t\bar{Z}_2 & Z_3 \end{pmatrix} : \begin{array}{l} Z_1 \; p \times p \text{ skew hermitian} \\ Z_3 \; q \times q \text{ skew hermitian} \end{array} \right\}$$

$\mathfrak{u}(n)$ ={all $n \times n$ skew hermitian matrices}

$\mathfrak{su}(p, q)$ ={all trace-free matrices in $\mathfrak{u}(p, q)$}

$\mathfrak{su}(n)$ ={all $n \times n$ trace-free skew hermitian matrices}

$\mathfrak{so}(n, \mathbb{C})$ ={all $n \times n$ skew symmetric complex matrices}

$$\mathfrak{so}(p, q) = \left\{ \begin{pmatrix} X_1 & X_2 \\ {}^tX_2 & X_3 \end{pmatrix} : \begin{array}{l} X_1 \; p \times p \text{ real skew symmetric} \\ X_3 \; q \times q \text{ real skew symmetric} \end{array} \right\}$$

$\mathfrak{so}(n)$ ={all $n \times n$ real skew symmetric matrices}

$$\mathfrak{sp}(n, \mathbb{C}) = \left\{ \begin{pmatrix} Z_1 & Z_2 \\ Z_3 & -{}^tZ_1 \end{pmatrix} : \begin{array}{l} Z_i \text{ complex} \\ Z_2, Z_3 \text{ symmetric} \end{array} \right\}$$

$$\mathfrak{sp}(n, \mathbb{R}) = \left\{ \begin{pmatrix} X_1 & X_2 \\ X_3 & -{}^tX_1 \end{pmatrix} : \begin{array}{l} X_i \text{ real} \\ X_2, X_3 \text{ symmetric} \end{array} \right\}$$

$$\mathfrak{sp}(n) = \left\{ \begin{pmatrix} Z_1 & Z_2 \\ -\bar{Z}_2 & \bar{Z}_1 \end{pmatrix} : \begin{array}{l} Z_1 \text{ skew hermitian} \\ Z_2 \text{ symmetric} \end{array} \right\}$$

Some low-dimensional Lie groups are familiar in physics, e.g. the special unitary and orthogonal groups. Further, $SL(2, \mathbb{Z})$ is a double cover of the Lorentz group $SO(1, 3)$, and $SU(2, 2)$ is a 4-fold cover of the conformal group in 4 dimensions. The second local isomorphism is of great importance in twistor theory.

Since a Lie algebra is a vector space, we can consider Lie algebras over various fields, in particular over $\mathbb{C}$ or $\mathbb{R}$. A real vector space V of even dimension is said to possess a *complex structure* if there is an $\mathbb{R}$–linear map

J of V to itself such that $J^2 = -I$, in which case it can be considered as a complex vector space $V^{\tilde{}}$ of half the dimension. In particular, the real 2-dimensional vector space $\mathbb{R}^2$ with points (x, y) can be considered as a complex 1-dimensional vector space $\mathbb{C}$ with points $z = x + iy$ by giving it a complex structure J corresponding to multiplication of z by i, which of course maps $\mathbb{R}^2$ to itself and satisfies $i^2 = -1$. On the other hand, a complex vector space E can always be considered as a real vector space $E_{\mathbb{R}}$ of double the dimension. We have $E = (E_{\mathbb{R}})^{\tilde{}}$.

For an arbitrary real vector space W, we can define a complex structure on $W \times W$ by $J : (X, Y) \mapsto (-Y, X)$. Then $(W \times W)^{\tilde{}}$ is called the *complexification* of W, denoted by $W_{\mathbb{C}}$. We have then the isomorphism $W_{\mathbb{C}} \cong (W \oplus JW)^{\tilde{}}$. A complex structure for a real Lie algebra $\mathfrak{g}$ is required to satisfy in addition

$$[X, JY] = J[X, Y], \quad \forall X, Y \in \mathfrak{g}.$$

Let $\mathfrak{g}$ be a complex Lie algebra. A *real form* of $\mathfrak{g}$ is a subalgebra $\mathfrak{g}_0$ of the real Lie algebra $\mathfrak{g}_{\mathbb{R}}$ such that as vector spaces they satisfy

$$\mathfrak{g}_{\mathbb{R}} = \mathfrak{g}_0 \oplus J\mathfrak{g}_0.$$

Thus $\mathfrak{g} \cong (\mathfrak{g}_0)_{\mathbb{C}}$, since $\mathfrak{g} \cong (\mathfrak{g}_{\mathbb{R}})^{\tilde{}} \cong (\mathfrak{g}_0 \oplus J\mathfrak{g}_0)^{\tilde{}} \cong (\mathfrak{g}_0)_{\mathbb{C}}$. A fundamental result (again due to Cartan) is that every semi-simple Lie algebra over $\mathbb{C}$ has a compact real form, where a real algebra $\mathfrak{g}$ is said to be *compact* if and only if there exists a compact Lie group G with Lie algebra isomorphic to $\mathfrak{g}$. Notice that $\mathfrak{g}$ may have non-isomorphic real forms, e.g. $\mathfrak{su}(2)$ and $\mathfrak{sl}(2, \mathbb{R})$ are both real forms of $\mathfrak{sl}(2, \mathbb{Z})$; but any two *compact* real forms are isomorphic.

We can now list (Table 3.2) all the simple Lie algebras $\mathfrak{g}$ over $\mathbb{C}$, with $G = $ a connected Lie group with Lie algebra $\mathfrak{g}_{\mathbb{R}}$, $K = $ a subgroup of G whose Lie algebra is a compact real form of $\mathfrak{g}$, $\widetilde{K} = $ the universal covering group of K, $Z(\widetilde{K}) = $ the centre of $\widetilde{K}$, and $n = $ rank of $\mathfrak{g}$.

The structures $\mathfrak{a}$, $\mathfrak{b}$, $\mathfrak{c}$, $\mathfrak{d}$ are called *classical*, and the five extra ones $\mathfrak{e}_6$, $\mathfrak{e}_7$, $\mathfrak{e}_8$, $\mathfrak{f}_4$, $\mathfrak{g}_2$ are called *exceptional*. The algebras have yet of course to be defined by giving, for example, their generators and their commutation relations, the classification of which, however, through Dynkin diagrams etc., would take far too long to go through here in detail. In any case, the classical structures should be obvious from what went before.

For the classical structures we give some further details in Table 3.3[4]

We note that in the case of $\mathfrak{d}_n$, for n odd, there exists one quotient of order 2 which is the well known $SO(2n)$; and for n even there are 3 quotients of order

---

[4]Note that both the symplectic group and its double cover are usually denoted by the same symbol $Sp(n)$. In mathematical physics the latter is sometimes called the *metaplectic* group.

$\mathfrak{g}$	K	$Z(\widetilde{K})$	dim K
$\mathfrak{a}_n$ $(n \geq 1)$	$SU(n+1)$	$\mathbb{Z}_{n+1}$	$n(n+2)$
$\mathfrak{b}_n$ $(n \geq 2)$	$SO(2n+1)$	$\mathbb{Z}_2$	$n(2n+1)$
$\mathfrak{c}_n$ $(n \geq 3)$	$Sp(n)$	$\mathbb{Z}_2$	$n(2n+1)$
$\mathfrak{d}_n$ $(n \geq 4)$	$SO(2n)$	$\mathbb{Z}_4$ n odd	$n(2n-1)$
		$\mathbb{Z}_2 \times \mathbb{Z}_2$ n even	
$\mathfrak{e}_6$	E_6	$\mathbb{Z}_3$	78
$\mathfrak{e}_7$	E_7	$\mathbb{Z}_2$	133
$\mathfrak{e}_8$	E_8	1	248
$\mathfrak{f}_4$	F_4	1	52
$\mathfrak{g}_2$	G_2	1	14

Table 3.2: Classical and exceptional structures

$\mathfrak{g}$	G	$\widetilde{K}$	$\widetilde{K}/Z(\widetilde{K})$
$\mathfrak{a}_n$	$SL(n+1,\mathbb{C})$	$SU(n+1)$	$SU(n+1)/\mathbb{Z}_{n+1}$
$\mathfrak{b}_n$	$SO(2n+1,\mathbb{C})$	$\mathrm{Spin}(2n+1)$	$SO(2n+1)$
$\mathfrak{c}_n$	$Sp(n,\mathbb{C})$	$Sp(n)$	$Sp(n)$
$\mathfrak{d}_n$	$SO(2n,\mathbb{C})$	$\mathrm{Spin}(2n)$	$PSO(2n)$

Table 3.3: Classical groups and algebras

2: the well known $SO(2n)$, and the two homeomorphic 'semi-spinor' groups. In the special case of $n = 4$, all three quotients are isomorphic by triality.

In low dimensions, we have the following isomorphisms:

$$\begin{aligned}
\mathfrak{a}_1 \cong \mathfrak{b}_1 \cong \mathfrak{c}_1 &\implies SU(2) \cong \text{Spin}(3) \cong Sp(1) \\
\mathfrak{b}_2 \cong \mathfrak{c}_2 &\implies \text{Spin}(5) \cong Sp(2) \\
\mathfrak{a}_3 \cong \mathfrak{d}_3 &\implies SU(4) \cong \text{Spin}(6) \\
\mathfrak{d}_2 \cong \mathfrak{a}_1 \oplus \mathfrak{a}_1 &\implies \text{Spin}(4) \cong SU(2) \times SU(2)
\end{aligned} \tag{3.2.4}$$

Now we come to the homotopy groups of Lie groups. We are concerned with only connected groups:

$$\pi_0(G) = 0.$$

A theorem of Weyl says that for G compact and semi-simple

$$\pi_1(G) = \text{ finite,}$$

implying that $\tilde{G}$ is still compact. In particular,

$$\begin{aligned}
\pi_1(SU(n)) &= 0 \\
\pi_1(SO(n)) &= \mathbb{Z}_2.
\end{aligned}$$

It is well known since Cartan that

$$\pi_2(G) = 0.$$

A result of Bott implies that for G compact and simple

$$\pi_3(G) = \mathbb{Z}.$$

For G compact, simply connected, and simple, we have

$$\pi_4(G) = 0 \text{ or } \mathbb{Z}_2.$$

The distinguishing criterion is somewhat long to state given our Spartan terminology.

Since the unitary groups $U(n)$ are topologically the product of $SU(n)$ with a circle S^1, their homotopy groups are easily computed using the product formula (3.1.1). We remind ourselves that $U(1)$ is topologically a circle and $SU(2)$ topologically S^3 (so see (3.1.2)).

Mainly using the homotopy sequence of fibre bundles (see Section 3.4) and the following coset relations:

$$\begin{aligned}
SO(n)/SO(n-1) &= S^{n-1} \\
SU(n)/SU(n-1) &= S^{2n-1} \\
Sp(n)/Sp(n-1) &= S^{4n-1} \\
\text{Spin}(7)/G_2 &= S^7 \\
\text{Spin}(9)/\text{Spin}(7) &= S^{15},
\end{aligned}$$

Isomorphism	Range
$\pi_i(SO(n)) \cong \pi_i(SO(m))$	$n, m \geq i + 2$
$\pi_i(SU(n)) \cong \pi_i(SU(m))$	$n, m \geq \frac{1}{2}(i + 1)$
$\pi_i(Sp(n)) \cong \pi_i(Sp(m))$	$n, m \geq \frac{1}{4}(i - 1)$
$\pi_i(G_2) \cong \pi_i(SO(7))$	$2 \leq i \leq 5$
$\pi_i(F_4) \cong \pi_i(SO(9))$	$2 \leq i \leq 6$
$\pi_i(SO(9)) \cong \pi_i(SO(7))$	$i \leq 13$

Table 3.4: Some isomorphisms for homotopy groups

we obtain the isomorphisms listed in Table 3.4. We have also the following general results:

$$\pi_6(SU(n)) = 0, \quad n \geq 4$$
$$\pi_6(SO(n)) = 0, \quad n \geq 5.$$

The homotopy groups of $SO(4)$ are readily computed from (3.2.4) using (3.1.1). Also from (3.1.3) and (3.2.4) we get

$$\pi_i(SO(3)) = \pi_i(SU(2)), \quad i \geq 2,$$
$$\pi_i(SO(5)) = \pi_i(Sp(2)), \quad i \geq 2,$$
$$\pi_i(SO(6)) = \pi_i(SU(4)), \quad i \geq 2.$$

Just for interest and to show the richness of the subject, we list a few other homotopy groups of some low-dimension unitary and orthogonal groups in Table 3.5.

3.3   Differential Forms

We shall endeavour in this section to give a taste of the beauty of de Rham theory and some related topics, which have a lot to do with the material to be dealt with in the following chapters. Our purpose here is mainly just to put the physical problems that we shall face in the proper mathematical context. The brief discussion is not meant to be even an introduction to the mathematical theory. We hope, however, that the physicist reader who wishes to know more may find such a synopsis helpful for familiarizing himself with the language before launching into abstract texts in algebraic topology; that is all.

Differential forms are perhaps easiest to understand through the two closely related topics of integration and differential equations.

	π_4	π_5	π_6	π_7	π_8	π_9	π_{10}
$SU(2)$	$\mathbb{Z}_2$	$\mathbb{Z}_2$	$\mathbb{Z}_{12}$	$\mathbb{Z}_2$	$\mathbb{Z}_2$	$\mathbb{Z}_3$	$\mathbb{Z}_{15}$
$SU(3)$	0	$\mathbb{Z}$	$\mathbb{Z}_6$	0	$\mathbb{Z}_{12}$	$\mathbb{Z}_3$	$\mathbb{Z}_{30}$
$SU(4)$	0	$\mathbb{Z}$	0	$\mathbb{Z}$	$\mathbb{Z}_{24}$	$\mathbb{Z}_2$	$\mathbb{Z}_{120}+\mathbb{Z}_2$
$SU(5)$	0	$\mathbb{Z}$	0	$\mathbb{Z}$	0	$\mathbb{Z}$	$\mathbb{Z}_{120}$
$SU(6)$	0	$\mathbb{Z}$	0	$\mathbb{Z}$	0	$\mathbb{Z}$	$\mathbb{Z}_3$
$SO(5)$	$\mathbb{Z}_2$	$\mathbb{Z}_2$	0	$\mathbb{Z}$	0	0	$\mathbb{Z}_{120}$
$SO(6)$	0	$\mathbb{Z}$	0	$\mathbb{Z}$	$\mathbb{Z}_{24}$	$\mathbb{Z}_2$	$\mathbb{Z}_{120}+\mathbb{Z}_2$
$SO(7)$	0	0	0	$\mathbb{Z}$	$\mathbb{Z}_2+\mathbb{Z}_2$	$\mathbb{Z}_2+\mathbb{Z}_2$	$\mathbb{Z}_8$
$SO(8)$	0	0	0	$\mathbb{Z}+\mathbb{Z}$	$\mathbb{Z}_2+\mathbb{Z}_2+\mathbb{Z}_2$	$\mathbb{Z}_2+\mathbb{Z}_2+\mathbb{Z}_2$	$\mathbb{Z}_{24}+\mathbb{Z}_{24}$
$SO(9)$	0	0	0	$\mathbb{Z}$	$\mathbb{Z}_2+\mathbb{Z}_2$	$\mathbb{Z}_2+\mathbb{Z}_2$	$\mathbb{Z}_8$
$SO(10)$	0	0	0	$\mathbb{Z}$	$\mathbb{Z}_2$	$\mathbb{Z}+\mathbb{Z}_2$	$\mathbb{Z}_{12}$

Table 3.5: Some homotopy groups for low $SU(n)$ and $SO(n)$

Let us take first the simplest case of the euclidean space $\mathbb{R}^n$, which is contractible, i.e. having all its homotopy groups trivial, and let $x^1, \ldots, x^n$ be its usual coordinates.

For reasons which will soon be apparent, we shall refer to functions of $x^1, \ldots, x^n$ also as *0-forms* in $\mathbb{R}^n$.

Next, a *1-form* is a covariant vector field on $\mathbb{R}^n$. Using $x^1, \ldots, x^n$, we can denote a basis for the space of 1-forms by $dx^1, \ldots, dx^n$, so that a general 1-form can be written as:

$$\theta = \theta_i \, dx^i. \tag{3.3.1}$$

For any 1-form θ, we can then define line integrals:

$$\int d\theta = \int \theta_i \, dx^i \tag{3.3.2}$$

along any curve in $\mathbb{R}^n$, as we did, for example, for the *gauge potential 1-form* $A_\mu dx^\mu$ in the preceding chapters. Further, given any function f of $x^1, \ldots, x^n$, or in other words, a 0-form in $\mathbb{R}^n$, its total differential:

$$df = \frac{\partial f}{\partial x^i} \, dx^i \tag{3.3.3}$$

is a 1-form, where we may thus regard the symbol d as representing a linear map from the space of 0-forms to that of 1-forms.

Just as 1-forms correspond to line integrals (3.3.2), higher forms can be defined to correspond to higher-dimensional integrals. For a 2-form, for example, the line elements dx^i above for 1-forms will be replaced by area elements, say $dx^i \wedge dx^j$. Now we know that in $\mathbb{R}^3$, for instance, an area can be assigned a direction, so that

$$\int f(x,y)\,dxdy = -\int f(x,y)\,dydx,$$

where to simplify the notation we have omitted the usual wedge product symbol $\wedge$ in the area elements $dx \wedge dy$ and $dy \wedge dx$. So it is natural to define 2-forms such that

$$dx^i dx^j = -dx^j dx^i, \qquad (3.3.4)$$

or equivalently to stipulate that the 1-forms dx^i are anticommuting. A general 2-form in $\mathbb{R}^n$ can then be written as:

$$\theta = \theta_{ij}\, dx^i dx^j \qquad (3.3.5)$$

in terms of $dx^i dx^j$ as basis, and integrals of θ can be defined over any 2-dimensional surface in $\mathbb{R}^n$, thus:

$$\int d\theta = \int \theta_{ij}\, dx^i dx^j. \qquad (3.3.6)$$

Again, given any 1-form (3.3.1), we can obtain a 2-form by:

$$d\theta = d\theta_i\, dx^i = \frac{\partial \theta_i}{\partial x^j}\, dx^j dx^i. \qquad (3.3.7)$$

More generally the totally skew expressions

$$dx^{i_1} \cdots dx^{i_k} \qquad (3.3.8)$$

form a basis for the space[5] of k-forms $\Omega^k(\mathbb{R}^n)$. On k-forms in $\mathbb{R}^n$, one can perform integrals over k-dimensional subsets of $\mathbb{R}^n$. Further, generalizing (3.3.3) we can define abstractly an operator d, called *exterior differentiation*, as a linear map:

$$d : \Omega^k(\mathbb{R}^n) \to \Omega^{k+1}(\mathbb{R}^n)$$

given by

$$d\left(adx^{i_1} \cdots dx^{i_k}\right) = dadx^{i_1} \cdots dx^{i_k}, \qquad (3.3.9)$$

where $da = (\partial a/\partial x^i)\, dx^i$. A fundamental property of d is

$$d^2 = 0. \qquad (3.3.10)$$

This follows simply from the symmetry of mixed partials:

$$d^2 f = \frac{\partial^2 f}{\partial x^i \partial x^j}\, dx^j dx^i \qquad (3.3.11)$$

---

[5]It is usual to denote this space by the letter Ω. We hope this will not be confused with the equally traditional use of the same letter to denote loop spaces.

which vanishes because $dx^j dx^i$ is skew.

Since we regard forms on $\mathbb{R}^n$ as expressions which can be integrated over subsets of $\mathbb{R}^n$, it is obvious that a k-form vanishes if $k > n$. Further, by means of integration, a k-form assigns to each k-dimensional subset of $\mathbb{R}^n$ a value of the integral. It can thus also be considered as a linear map of rank k skew tensor fields to numbers. In particular, a 1-form maps any vector field on $\mathbb{R}^n$ to a number. A familiar example in gauge theory is the potential 1-form which assigns to each vector at x, say $\{dx^\mu\}$, a value for the phase change under parallel transport to $x + dx^\mu$, which in usual physicists' notation is given just as $A_\mu(x)dx^\mu$. In abelian theories, $A_\mu dx^\mu$ is a real number, but in nonabelian theories, it takes values in the gauge Lie algebra. In the latter case, the potential 1-form is said to be Lie algebra-valued. These observations will be of use later in Section 3.4 for defining connections on fibre bundles.

Example 1: Vector calculus on $\mathbb{R}^3$

The exterior differentiation is the usual gradient, curl or divergence, according to the degree of the form on which it acts:

d (0-forms) = gradient
d (1-forms) = curl
d (2-forms) = divergence

Thus the well-known identities curl grad = 0 and div curl = 0 of vector calculus are both special cases of (3.3.10). Notice that since $\Omega^0(\mathbb{R}^3)$, $\Omega^3(\mathbb{R}^3)$ are 1-dimensional, and $\Omega^1(\mathbb{R}^3)$, $\Omega^2(\mathbb{R}^3)$ are 3-dimensional, we have made certain natural identifications above, i.e. volumes as scalars and areas as vectors.

In case $\mathbb{R}^N$ is endowed with a metric, or more generally in the case of a Riemannian manifold X, then to every form θ one can define a dual form $^*\theta$ by means of the *Hodge star* operation, which is an isomorphism:

$$* : \Lambda^p \xrightarrow{\simeq} \Lambda^{n-p}.$$

In tensor language, where at every point in X, a p-form is a rank p skew-symmetric tensor, this amounts to raising all the indices of the tensor by the metric and then contracting with the ϵ-tensor. For example, in Minkowski space, the dual of the electromagnetic tensor is given by (2.2.4):

$$^*f_{\mu\nu} = -\tfrac{1}{2}\epsilon_{\mu\nu\rho\sigma}f^{\rho\sigma}.$$

Poincaré lemma

In elementary differential equation theory, we are familiar with the following facts. A total differential can be integrated along any curve between its end points a and b (Fundamental Theorem of Calculus) giving:

$$\int_a^b df = f(b) - f(a). \qquad (3.3.12)$$

On the other hand, a general 1-form (3.3.1) need not be integrable in this manner. The integrability condition

$$\frac{\partial \theta_i}{\partial x^j} - \frac{\partial \theta_j}{\partial x^i} = 0 \qquad (3.3.13)$$

is often written in elementary differential equation contexts as

$$d\theta = 0, \qquad (3.3.14)$$

using the same notation as (3.3.7) and (3.3.9). This means that in this case we can find a function f whose total differential is θ:

$$\theta = df. \qquad (3.3.15)$$

In general, any k-form θ satisfying $d\theta = 0$ (3.3.14) is said to be *closed*; it is *exact* if it can be expressed as the differential of another form, i.e. $\theta = df$ (3.3.15). This is essentially the same terminology as for differential equations. Because the operator d squares to zero (3.3.10), an exact form is always closed. In $\mathbb{R}^n$, the converse is also true. This result is known as the

Poincaré Lemma   *In $\mathbb{R}^n$, a form is exact if and only if it is closed.*

Let us immediately point out that the Poincaré lemma as stated above does not hold globally for a general manifold. However, even in this case (3.3.15) can always be solved 'locally', i.e. in each coordinate patch (which is homeomorphic to $\mathbb{R}^n$). The question is whether these solutions agree on the overlaps. We shall return to discuss this question later.

For gauge theories, the Poincaré lemma is relevant in telling us whether a given field tensor can or cannot be derived from a gauge potential. Example 2 below will give us some indication of the importance of this result for Maxwell theory, the physical significance of which will be discussed again in Chapter 4. Similar physical implications also exist for nonabelian gauge theories, but this will require a (nontrivial) extension of the Poincaré lemma which will be described in Section 4.5.

In (3.3.12) we have seen that the line integral of an exact 1-form along any curve can be evaluated in terms of its end-points. In elementary vector calculus, we are also familiar with Stokes' and Gauss' theorems which give integrals over surfaces and volumes of the curl and divergence of vectors, both exact forms in the present language, as integrals of the vectors over the corresponding boundaries. Using now the concise language of differential forms, we can incorporate all those useful 'integration laws' of vector calculus into a single statement. If we call the *support* of a function the smallest closed set outside of which the function vanishes, then we have:

Stokes' Theorem   *Let X be an oriented n-dimensional manifold with boundary ∂X, and ω an $(n-1)$-form with compact support. Then*

$$\int_X d\omega = \int_{\partial X} \omega,$$

where ∂X is equipped with the induced orientation.

de Rham cohomology

Since exact forms are automatically closed but, for a general manifold, not all closed forms are exact, we might want to enquire about those closed forms which are not exact. This is one question addressed in de Rham cohomology.

The pth de Rham cohomology group is defined as

$$H^p = \{\text{closed } p\text{-forms}\}/\{\text{exact } p\text{-forms}\}. \qquad (3.3.16)$$

One can easily check that H^p is not only an abelian group but also a vector space. It is a measure of the departure from the Poincaré lemma and in the case of $p = 1$ gives the 'interesting' solutions to (3.3.14).

The cohomology groups are topological invariants of the space, and in terms of them we can restate the Poincaré lemma as:

$$H^p(\mathbb{R}^n) = \begin{cases} \mathbb{R} & p = 0, \\ 0 & p \neq 0. \end{cases}$$

where the first statement for $p = 0$ simply comes from the fact that closed 0-forms are the constant functions.

The Poincaré lemma tells us that the cohomology of $\mathbb{R}^n$ is not very interesting. However, the concepts of differential forms and de Rham cohomology go over easily to any differentiable manifold. Recall that roughly speaking a manifold X is given by a collection of open sets U, each diffeomorphic to $\mathbb{R}^n$, with differentiable patching conditions on overlaps. So one can define a differential form ω on X by a collection of differential forms ω_U for each U in such a way that they are compatible on overlaps. In this way one can build up a de Rham theory, in which the cohomology spaces $H^p(X)$ are again topological invariants (and not necessarily trivial). Indeed, for spheres S^n and Riemann surfaces X_g with g holes, we have:

$$H^p(S^n) = \begin{cases} \mathbb{R} & p = 0 \text{ or } n, \\ 0 & \text{otherwise.} \end{cases}$$

$$H^p(X_g) = \begin{cases} \mathbb{R} & p = 0 \text{ or } 2, \\ \mathbb{R}^{2g} & p = 1, \\ 0 & \text{otherwise.} \end{cases}$$

A similar construction also works for manifolds with boundary, where some of the sets U are diffeomorphic to upper half spaces $\mathbb{H}^n$ rather than $\mathbb{R}^n$, where $\mathbb{H}^n = \{(x_1, \ldots, x_n) : x_n \geq 0\}$.

For most reasonable manifolds[6] (and these include compact manifolds), the de Rham cohomology spaces $H^p(X)$ are finite-dimensional. Moreover, the wedge product of two differential forms induces a product between cohomology spaces called the *cup product*.

Example 2: Maxwell theory

In free Maxwell theory (no source, no monopole), the potential a (which is a 1-form) exists everywhere in $\mathbb{R}^4$ and the field strength is a 2-form given by

$$f = da, \tag{3.3.17}$$

where here d plays the role of a curl as indicated in Example 1. It then follows by (3.3.10) that

$$df = 0. \tag{3.3.18}$$

These two equations should be compared with (3.4.4) and (3.4.9) of the next section. Now (3.3.18) is half of the Maxwell equations and the other half is

$$d^*f = 0. \tag{3.3.19}$$

---

[6]Technically, the existence of a finite good cover is a sufficient condition.

By the Poincaré lemma we can deduce the existence of what we can call (see later in Section 5.1) the dual potential $\bar{a}$ such that

$$^*f = d\bar{a}. \tag{3.3.20}$$

In the presence of a point source the equation (3.3.18) still holds on $\mathbb{R}^4$ as before, but (3.3.19) holds only on $\mathbb{R}^4 - \mathbb{R}_t$, where $\mathbb{R}_t$ is the world-line of the source. The de Rham cohomology of this space can be computed:

$$H^p(\mathbb{R}^4 - \mathbb{R}_t) = \begin{cases} \mathbb{R} & p = 0 \text{ or } 2, \\ 0 & \text{otherwise.} \end{cases}$$

It can be shown that in this case, the cohomology class of $^*f$ is non-zero, and can be interpreted as the (unquantised) charge of the source.

Dually, if we have a monopole instead, then the conclusions regarding f and $^*f$ are interchanged. We notice that the monopole charge is also unquantised, as the gauge group here is tacitly taken to be $\mathbb{R}$ and not $U(1)$.

From Stokes' theorem, it follows that integrals of exact forms over compact manifolds with no boundaries all vanish, and since by definition forms within a cohomology class differ only by exact forms, it follows further that such integrals of forms within the same cohomology class have all the same value and can be considered as the integral for the whole class. Further, if we define H_c^p to be the corresponding cohomology spaces for forms with compact support, then for an n-dimensional manifold X we get a *pairing*, defined by:

$$H^p(X) \otimes H_c^{n-p}(X) \longrightarrow \mathbb{R}$$
$$(\omega, \eta) \longmapsto \int_X \omega\eta$$

mapping pairs of cohomology classes to a real number, somewhat analogous to the inner product between a vector and its dual in a vector space. We can therefore define also a duality between cohomology classes thus:

$$H^p(X) \simeq (H_c^{n-p}(X))^*.$$

which is known as *Poincaré duality*. (This has, of course, little to do with the duality between forms introduced above via the Hodge star.) When X is compact, we have more simply

$$H^p(X) \simeq (H^{n-p}(X))^*.$$

At the level of connected components and functions on spaces, there is a kind of duality between homotopy (π_0) and cohomology (H^0). This duality, however, does not easily extend to higher dimensions. What one does define are the homology groups H_p of spaces, and these are, generally speaking, dual to cohomology. In particular, for compact oriented manifolds X there is exact duality, and hence

$$H^p(X) \simeq H_p(X),$$

so that for some purposes one can treat them as identical. An example is the above statement of Poincaré duality, which is more usually stated as a pairing between cohomology and homology. Unfortunately homotopy is more closely related than homology to geometry and so to physics. The relation between the two is in general quite complicated. However, when X is simply connected, let $\pi_n(X)$ be the first non-vanishing homotopy group. Then the Hurewicz isomorphism theorem says that $H_q(X) = 0$, $\forall\ 0 < q < n$, and $H_n(X) = \pi_n(X)$. This suggests that for many purposes it is enough to compute the homology instead of the homotopy groups.

Apart from its many beautiful features, the power and usefulness of cohomology theory come from the fact that the H^p can be calculated using standard techniques because relations between the manifolds themselves translate into relations between their cohomologies. For the treatment of this problem, together with that in the related homology theory, a general scheme has been developed using a machinery which those readers who are unacquainted with the language may find difficult to unravel. The following are some concepts and terms figuring conspicuously in mathematical texts which the reader may find useful.

A sequence of linear maps of vector spaces V_i

$$\cdots \longrightarrow V_{i-1} \xrightarrow{f_{i-1}} V_i \xrightarrow{f_i} V_{i+1} \longrightarrow \cdots$$

is called a *complex* if $f_i \circ f_{i-1} = 0$ for all i along the sequence, where $\circ$ denotes the composition of two maps, in this case it means first operating with f_{i-1} then with f_i. The sequence is said to be *exact* if

$$\operatorname{im} f_{i-1} = \ker f_i,$$

where $\operatorname{im} f_{i-1}$ denotes the image of the map f_{i-1}, i.e. $f_{i-1}(V_{i-1})$ in V_i, and $\ker f_i$ the kernel of f_i, namely the subspace of V_i which is mapped to 0 by f_i. Clearly exactness is the stronger of the two conditions. A natural example of a complex is the de Rham *differential complex*:

$$\cdots \longrightarrow \Omega^{i-1} \xrightarrow{d} \Omega^i \xrightarrow{d} \Omega^{i+1} \longrightarrow \cdots \qquad (3.3.21)$$

Poincaré lemma tells us that for $X = \mathbb{R}^n$, the following differential complex is in fact exact:

$$0 \longrightarrow \mathbb{R} \xrightarrow{k} \Omega^0 \xrightarrow{d} \Omega^1 \xrightarrow{d} \cdots$$

where the first map k simply takes a real number to the constant function taking that value. An exact sequence of the form

$$0 \longrightarrow A \longrightarrow B \longrightarrow C \longrightarrow 0$$

is called a *short exact sequence*. Here exactness at A means that the only element of A that vanishes under the map $A \to B$ is 0, which is the whole image of the map $0 \to A$. This is equivalent to the map $A \to B$ being injective. Exactness at C means that the map $B \to C$ is surjective, since its image is all of C, every element of which must vanish under the map $C \to 0$.

A differential complex (3.3.21) can be denoted by a single letter:

$$\Omega = \bigoplus_{p \in \mathbb{Z}^+} \Omega^p.$$

The useful result is that a short exact sequence of differential complexes:

$$0 \longrightarrow A \longrightarrow B \longrightarrow C \longrightarrow 0$$

induces a long exact sequence on the corresponding cohomology groups. The Mayer–Vietoris sequence, a basic technique for computing H^p, is such an example. If enough H^p vanish, one can then deduce injectivity, surjectivity or even bijectivity among the remaining H^p, as in the discussion above. A concrete example for the application of the exact sequence technique to a physics problem will be found in Section 3.4 in the subsection on reduction of bundles.

Example 3: Mayer–Vietoris sequence

Let $X = U \cup V$, where U and V are open. Then the short exact sequence of differential forms

$$0 \to \Omega(X) \to \Omega(U) \oplus \Omega(V) \to \Omega(U \cap V) \to 0$$

induces a long exact sequence in cohomology

$$\cdots \to H^{p-1}(U \cap V) \to H^p(X) \to H^p(U) \oplus H^p(V) \to H^p(U \cap V) \to \cdots$$

If we take U and V as coordinate patches, i.e. homeomorphic images of $\mathbb{R}^n$, then by the Poincaré lemma the direct sum terms in the above sequence vanish, and the properties of $H^p(X)$ can be deduced from those of the cohomologies of the overlap $U \cap V$.

In the treatment above, we have taken perhaps an extremely simplistic approach. There are many other ways to define cohomology (e.g. simplicial, singular, Čech, etc.) for spaces far more general than the ones we are interested in. So there are also for homology. They fit into very general schemes of definitions and computations that bear names like exact sequences, complexes, double complexes, categories, functors, etc. some of which have been introduced above. Though sometimes unfortunately obscured to the uninitiated by what is known in the trade as *abstract nonsense*, this machinery treats all the different aspects of the whole subject together and does the computations in a systematic way — and also lends itself to generalizations. Surprisingly, one such generalization, though seemingly abstract, turns out to have something to do with physics and will be briefly discussed below.

Non-commutative or extended geometry

The algebra $C(X)$ of continuous complex-valued functions on a space X is commutative. If X is compact, then $C(X)$ can be made into a C^*-algebra[7]; if X is only locally compact, then to obtain a C^*-algebra we have to restrict to the algebra $C_0(X)$ of continuous functions vanishing at infinity. A well-known theorem due to Gel'fand gives the converse. If $\mathfrak{A}$ is a commutative C^*-algebra, the set of non-zero homomorphisms of $\mathfrak{A}$ into $\mathbb{C}$, called its *spectrum* $\hat{\mathfrak{A}}$, is a locally compact space when considered as a subset of the dual of $\mathfrak{A}$. Then $\hat{\mathfrak{A}}$ can be identified with X, and moreover $C(\hat{\mathfrak{A}}) = \mathfrak{A}$. (Here we have glossed over the difference between $C(X)$ and $C_0(X)$.) It is easy to see heuristically what is going on. An element $t \in \hat{\mathfrak{A}}$ is a map $\mathfrak{A} \to \mathbb{C}$. The identification of $t \in \hat{\mathfrak{A}}$ with an element $x \in X$ is given by equating the complex numbers $t(f) = f(x)$, $\forall f \in \mathfrak{A}$. To construct X explicitly from $\mathfrak{A}$, one notices that an element $t \in \hat{\mathfrak{A}}$ can be identified with its kernel, which is a maximal ideal of $\mathfrak{A}$. Going back to functions, we know that a maximal ideal is a subset of functions which vanish at a certain point, so that a maximal ideal defines and is defined by a point in the original space. Hence X can be recovered from the maximal ideals $\ker t$, $t \in \hat{\mathfrak{A}}$. In this way, the Gel'fand correspondence completely classifies all commutative C^*-algebras.

A *derivation* of an algebra $\mathfrak{A}$ is a map from $\mathfrak{A}$ into itself satisfying the Leibnitz rule on products. When $\mathfrak{A} = C(X)$ (or rather, a dense subalgebra of C^∞ functions, but in this rough introduction we shall neglect such analytic niceties), then a derivation δ on $\mathfrak{A}$ corresponds to a vector field on X. In local

---

[7]A concrete definition of a C^*-algebra is as a subalgebra of the algebra of all bounded linear operators on a Hilbert space which is closed under involution (complex conjugation in our case) and closed in the operator norm.

coordinates,

$$\delta = f_k(x) \frac{\partial}{\partial x^k},$$

where f_k are continuous functions. Conversely, every vector field on X defines a derivation on $C(X)$.

Within operator algebras there is no reason to consider only *commutative* C^*-algebras. What is more, in physics we have learnt to go from commutative structures to non-commutative ones, e.g. quantum mechanics, Yang–Mills theory. Gel'fand's theorem tells us that non-commutative C^*-algebras cannot be the function algebras on spaces with points. The question is: can they be function algebras for more general 'pointless' spaces? More interesting still: can we do 'differential geometry' in these algebras? We know that we can do much of differential and algebraic geometry on the algebra $C(X)$ without reference to X; so the above questions are mathematically reasonable. Whether they make sense physically will be discussed in Chapter 6.

Just as one can build up the de Rham differential complex $\Omega(X)$ from the algebra of functions of the manifold X, there are general constructions for non-commutative differential forms given any algebra, which lead ultimately to formulae for calculating the index of 'spaces'. However, for our present purposes, we shall restrict ourselves to a simple special case of Connes' non-commutative geometry where there is a group action. Let $\mathfrak{A}$ be a (non-commutative) C^*-algebra, G a locally compact group and δ a representation of G into the Lie algebra of derivations of $\mathfrak{A}$. Then the non-commutative analogue of the de Rham complex is the complex of left-invariant forms on G with values in $\mathfrak{A}$. Left-invariant forms are those which are invariant under left multiplication by an element of G. The product in this $\Omega(\mathfrak{A})$ is given by

$$(a_1 \otimes \omega_1)(a_2 \otimes \omega_2) = a_1 a_2 \otimes \omega_1 \omega_2,$$

where $a_i \in \mathfrak{A}$, $\omega_i \in \Omega(G)$, $i = 1, 2$. The exterior derivative d is such that:

1. $da\,(V) = \delta_V(a)$, for $a \in \mathfrak{A}$, $V \in \mathfrak{g}$,

2. $d(\tilde\omega_1\,\tilde\omega_2) = d\tilde\omega_1\,\tilde\omega_2 + (-1)^{|\tilde\omega_1|}\,\tilde\omega_1\,d\tilde\omega_2$, $\forall \tilde\omega_1,\ \tilde\omega_2 \in \Omega(\mathfrak{A})$, where $|\tilde\omega_1| = $ degree of $\tilde\omega_1$,

3. $d^2\tilde\omega = 0$, $\forall\ \tilde\omega \in \Omega(\mathfrak{A})$.

Our ultimate aim here is to generalize the gauge theory concept where the gauge symmetry is attached to extended objects such as strings and not just points. We find the above construction gives us an appropriate setting for such a purpose, but before we go on we must study the mathematical structure of the usual (pointwise) gauge theory — fibre bundles.

3.4   Fibre Bundles

Descriptively, a fibre bundle is a twisted product of two spaces, say X and F, where F is acted on by a group G, and the twist in the product has been effected by the group action. In Figure 2.8, we have already seen two illustrations of fibre bundles in the Möbius band and the gauge field configuration of the monopole. To give a more precise definition, we shall call X the *base space*, F the *fibre* on which the *structure group* G acts effectively (i.e. $gy = y \ \forall y \in F \Rightarrow g =$identity). We have then a *fibre bundle* over X with *total space E*, whenever we are given a *projection*

$$\pi : E \to X,$$

and an open cover $\{U_\alpha\}$ for X such that for each U_α there exists a homeomorphism:

$$\phi_\alpha : U_\alpha \times F \to \pi^{-1}(U_\alpha)$$

satisfying the following three conditions:

(i) $\pi \circ \phi_\alpha =$ identity on U_α,

(ii) if we define

$$\phi_{\alpha,x} : \quad F \quad \longrightarrow \quad \pi^{-1}(x)$$
$$y \quad \longmapsto \quad \phi_\alpha(x,y)$$

then on the overlap $U_\alpha \cap U_\beta$, the composite

$$\phi_{\beta,x}^{-1} \, \phi_{\alpha,x} : F \to F$$

coincides with the action of a (unique) element of G,

(iii) the map

$$\phi_{\alpha\beta} : U_\alpha \cap U_\beta \longrightarrow G$$

that sends $x \in V_\alpha \cap V_\beta$ to this element of G is continuous.

When we are interested in differentiable manifolds then of course continuity of the maps will be replaced by smoothness.

A convenient picture of a fibre bundle is:

$$F \xrightarrow{\ i\ } E$$
$$\downarrow \pi$$
$$X$$

where i is the natural map of the fibre into the total space. For $x \in X$, $\pi^{-1}(x)$ is called the *fibre above* x and is a homeomorphic copy of F.

A fibre bundle is said to be *trivial* when it is homeomorphic to just a straightforward Cartesian product $X \times F$, like for instance the strip in Figure 2.8b, and the group action is trivial. A well-known example of a non-trivial bundle is of course the Möbius band, with $X = S^1$ the circle, F the line segment represented by the closed interval $[0, 1]$, and $G = \mathbb{Z}_2$, namely up-down flips of the line segment F. Notice that the group action has to be specified in the definition. For example, if we replace the line segment above by a circle, we get a nontrivial 'twisted torus'. However, if instead of $\mathbb{Z}_2$ we allow all rotations of the circle (i.e. the group becomes $U(1)$) then the 'twisted torus' becomes trivial.

Two types of fibre bundles are of particular interest: principal bundles and vector bundles. A *principal bundle* has $F = G$ and the group action is left multiplication. A *vector bundle* has $F = V$ an n-dimensional vector space and $G = GL(n, \mathbb{R})$ or $GL(n, \mathbb{C})$.

Given an arbitrary bundle $E \xrightarrow{\pi} X$ with structure group G, one can construct its *associated principal bundle* using the same data above except replacing the fibre F by G and letting G act on itself by left multiplication. More generally we say that two bundles are *associated* if their corresponding associated principal bundles are isomorphic. Suppose we have a smooth map $f : X \to \tilde{X}$, and a principal bundle $\tilde{\pi} : \tilde{P} \to \tilde{X}$, then we can build a principal bundle $\pi : P \to X$ and a map $\phi : P \to \tilde{P}$ in such a way that the following diagram of maps commute:

$$\begin{array}{ccc} P & \xrightarrow{\phi} & \tilde{P} \\ \pi \downarrow & & \downarrow \tilde{\pi} \\ X & \xrightarrow{f} & \tilde{X} \end{array}$$

i.e. $\tilde{\pi} \circ \phi = f \circ \pi$. For the fibre above $x \in X$ we just take the fibre above $f(x) \in \tilde{X}$, and the total space P is obtained by putting these fibres side by side, i.e.

$$P = \bigcup_{x \in X} (\{x\} \times \tilde{\pi}^{-1}(f(x))).$$

A point in P is then given by (x, z), where $z \in \tilde{\pi}^{-1}(f(x))$. The projection π is simply defined by $\pi(x, z) = x$. The right action by G is given by

$$(x, z)\, g = (x, zg), \quad g \in G.$$

The map $\phi : P \to \tilde{P}$ is then $\phi(x, z) = z$. The principal bundle P so constructed is called the *pull-back* of $\tilde{P}$ and is denoted by $f^*\tilde{P}$. Similarly one can construct the pull-back of a vector bundle.

A map $s : X \to E$ such that when composed with the projection π gives the identity on X is called a *section*. The existence of sections is an important question. They always exist on trivial bundles. A principal bundle admits a section if and only if it is trivial. A vector bundle always has at least one section, namely the zero section consisting of the 'origins' of the fibres.

Examples

1. **Covering spaces.** Here G is discrete and is a quotient group of the fundamental group of the base space X. For instance, $SU(2)$ is a principal bundle over $SO(3)$ with group $\mathbb{Z}_2$. This can be seen easily as the space of $SU(2)$ is S^3 and the space of $SO(3)$ is obtained from S^3 by antipodal identification (see also Sections 3.1, 3.2 and 5.2).

2. **Coset spaces.** Let H be a subgroup of G, then G is a principal bundle over the coset space G/H with group H.

3. **Tangent bundles.** Let X be a manifold, then one can construct a vector bundle where a fibre above $x \in X$ is the tangent space to X at that point x. The group is $GL(n)$, where n is the dimension of X. If X happens to be a Riemannian manifold, then it is usual to restrict to $SO(n)$, a subgroup of $GL(n)$, as it is then more meaningful to consider only those transformations which preserve the Riemannian metric. If we replace the tangent spaces by various other spaces of tensors, then we get the corresponding tensor bundles. Furthermore, in this language a vector field is a section of the tangent bundle, and a 1-form is a section of the cotangent bundle.

4. **Frame bundles.** Let X be an n-dimensional manifold. A *frame u* at $x \in X$ is an ordered basis of the tangent space at x. The space of all frames at x can thus easily be identified with the general linear group $GL(n, \mathbb{R})$. The total space of the frame bundle is the collection of all frames at all points $x \in X$, and the bundle is thus a principal bundle with group $GL(n, \mathbb{R})$. A section of this bundle is called a *moving frame*. If X is an oriented Riemannian manifold, then one can restrict to orthonormal frames. The structure group is then reduced to $SO(n)$. Again, if X is a complex manifold with a hermitian structure defined in its tangent bundle, then one can restrict to unitary frames and the structure group is then reduced to $U(N)$. The concept of reduction to a subgroup of the structure group will be treated more fully later.

In the theory of monopoles, we are particularly interested in principal bundles over S^2. In fact, if we consider any point-like object, whether source or monopole, then the relevant flat space-time is $\mathbb{R}^4$ with the world-line of this point-like object removed. The spatial part is thus $\mathbb{R}^3 - \{point\} \approx S^2 \times \mathbb{R}^+$, which has the same topology as S^2 (see also Section 2.3 for more detail). Multiplying by the time axis does not change the topology.

By using the *homotopy sequence* of a principal bundle (see later in the subsection on reductions of bundles), especially the homomorphism

$$\pi_p(X) \longrightarrow \pi_{p-1}(G)$$

Steenrod and Pontrjagin were able to give the complete classification of bundles over a sphere S^p.

Theorem *If G is connected, then the set of equivalence classes of fibre bundles over S^p with group G is in 1–1 correspondence with $\pi_{p-1}(G)$.*

Since the gauge field configuration in the presence of a monopole is that of a principal G-bundle over S^2, we obtain the important result of Chapter 2, namely that monopole charges, or equivalent classes of such bundles, are in 1–1 correspondence with elements of $\pi_1(G)$, or in other words, homotopy classes of closed curves on G.

When $G = U(1)$ in the electromagnetic case, the charges are integers and coincide with the Chern class of the corresponding bundle (see the subsection on characteristic classes below). Further, the total space of the $U(1)$-bundle over S^2 corresponding to monopole charge 1 happens to have the same topology as the 3-dimensional sphere S^3, and as such is called the *Hopf bundle*:

$$S^1 \to S^3 \to S^2.$$

This is a particular case of fiberings of spheres by spheres, since the fibre $U(1)$ has the topology of the circle S^1. The two other well-known cases are

$$S^3 \to S^7 \to S^4$$

where the group $SU(2)$ has the topology of S^3, and

$$S^7 \to S^{15} \to S^8$$

which is not a principal bundle.

The $U(1)$ bundle corresponding to monopole charge 2 has the non-orientable real projective 3-space $\mathbb{RP}^3$ as its total space. It can be obtained from S^3 by identifying antipodal points. Higher monopole charges m correspond to m-fold identification of points on S^3 which are known as *lens spaces*. As expected, different monopole charges correspond to total spaces of different topologies, with different fundamental groups $\mathbb{Z}_m$.

Connections

The idea of a connection is to enable us to parallelly transport geometrical objects. Although historically the first considerations on connections
were about parallel transport of tangent vectors around manifolds (Christof-
fel symbols, etc.), the concept has since been generalised to a wide variety of
contexts, and because of our interest in gauge theory, we shall discuss here
mainly connections only on principal bundles.

In gauge theories, we are interested in parallel transport of 'phases' from
point to neighbouring point in space-time. As explained in the preceding
chapters, this is represented by the gauge potential which, in the language of
differential forms of Section 3.3 and of fibre bundles here, is a 1-form $A_\mu dx^\mu$
on the base space X. However, because the bundle may be nontrivial, e.g. in
the presence of a monopole, the gauge potential may have to be patched. In
that case, it is not just a single 1-form on X, but a family of 1-forms each
defined on a patch of X, and the whole family has to be related by patching
conditions in the overlapping regions, as explained in Section 2.3. Because of
this complication, mathematically connections are more properly defined as
1-forms on the total space of the bundle. We shall examine below some such
definitions of connections and clarify their relationship with the language of
gauge potentials as adopted by physicists.

First we need to introduce some useful concepts.

(a) Given a map $f : X \to X'$ we can define its differential f_* at $x \in X$,
a linear map from the tangent space $T_x X$ to the tangent space $T_{f(x)} X'$, as
follows. Given a tangent vector V at x, choose any curve $x(t)$ in X such that
$x(0) = x$ and V is tangent to $x(t)$ at x. Then the image $f_* V$ is the tangent
vector to the image curve in X' at $f(x)$. It can be shown that the construction
is independent of the curve chosen. Similarly, recalling from Section 3.3 that
1-forms map vector fields to numbers, for any 1-form ω' on X' one can define
$f^* \omega'$ by

$$(f^* \omega') V = \omega' (f_* V),$$

where V is now a vector field on X. Notice that although $f_* V$ may not be a
vector field in general[8], $f^* \omega'$ is always a 1-form.

(b) A group G acts on itself by left translation, so that for each $a \in G$, we
have the map

$$L_a : g \mapsto ag.$$

A vector field V on G is said to be *left invariant* if $(L_a)_* V = V$, $\forall g \in G$. Thus,
a more usual definition than the one given in Section 3.2 of the Lie algebra $\mathfrak{g}$
of G is as the set of left invariant vector fields on G. Once we have invariance

---

[8]as tangent vectors at two different points may be mapped to tangents at the same point

by left translation, it is quite easy to go from this definition to that previously given as the tangent space at e, and vice versa.

(c) Suppose now a group G acts on the right on a manifold X. Then for each $A \in \mathfrak{g}$, the action induces a vector field $\sigma(A)$ on X as follows. At each $x \in X$, consider the action of the 1-parameter subgroup $\exp tA$, whose orbit is a curve in X passing through x at $t = 0$. The tangent to this curve at x is the required vector. We call $\sigma(A)$ the *fundamental vector field* corresponding to $A \in \mathfrak{g}$.

Return now to the definition of a connection on a bundle. One version of this goes as follows. Let P be a principal bundle over X, with structure group G. G acts on P on the right as follows. Given $a \in G$, $u \in P$, we have:

$$R_a(u) = \phi_\alpha(x, (\phi_{\alpha,x}^{-1}(u))a), \qquad (3.4.1)$$

where $x = \pi(u) \in U_\alpha$, and where we recall that $\phi_{\alpha,x}^{-1}(u)$ is an element of G. Notice that this action moves points along the same fibre, and it is a right action because $R_{a_1 a_2}(u) = R_{a_2}(R_{a_1}(u))$. A *connection 1-form* ω on P is a $\mathfrak{g}$-valued 1-form on P satisfying:

(1) $\omega(\sigma(A)) = A, \ \forall A \in \mathfrak{g}$,

(2) $\omega((R_a)_* V) = \text{ad}\,(a^{-1})\omega(V), \ \forall a \in G$, and for every vector field V on P.

where the adjoint action $\text{ad}\,(a^{-1})$ of a^{-1} on $A \in \mathfrak{g}$ is often written as $a^{-1}Aa$.

Another, more descriptive, way to define a connection is to say that it specifies a horizontal subspace $\mathcal{H}_u$ of the tangent space T_u at each point $u \in P$, in such a way that:

(1') $T_u = \mathcal{H}_u \oplus \mathcal{V}_{u,,}$

(2') $\mathcal{H}_{R_a(u)} = (R_a)_* \mathcal{H}_u, \ a \in G$,

(3') If V is a differentiable vector field on P, then so are its vertical and horizontal components.

Here $\mathcal{V}_u$, the vertical subspace, consists of those tangent vectors which are tangent to the fibre at u. The map σ above induces a linear isomorphism of $\mathfrak{g}$ onto $\mathcal{V}_u$.

It is not hard to go from one of the above two descriptions to the other. Given a decomposition (1') of the tangent space T_u, the 1-form ω is defined by assigning to each tangent vector at u its vertical component, which by the isomorphism induced by σ, is identified with an element of $\mathfrak{g}$. Conversely, given the 1-form ω, the horizontal subspace consists of those tangent vectors at u which are annihilated by ω. Using a basis $\{A_1, \ldots, A_k\}$ for $\mathfrak{g}$, the 1-form ω

can be written as $\omega = \sum_{j=1}^{k} \omega^j A_j$, where ω^j are just ordinary (i.e. real-valued) 1-forms on P. Then the vertical component v of V is given by:

$$v(V) = \sum_{j=1}^{k} \omega^j(V)\sigma(A_j).$$

Next, we shall show that a connection as defined above on a principal bundle can indeed also be represented by a *family* of 1-forms on the base space X, as envisaged usually by physicists. Let $\{U_\alpha\}$ be the cover of X used in defining the principal bundle P, with homeomorphisms

$$\phi_\alpha : U_\alpha \times G \longrightarrow \pi^{-1}(U_\alpha)$$

and transition functions

$$\phi_{\alpha\beta} : U_\alpha \cap U_\beta \longrightarrow G.$$

Above the patch U_α, we have the section s_α defined by

$$s_\alpha(x) = \phi_\alpha(x, e),$$

where e is the identity element of G. Then the 'local' connection 1-forms are defined by $\omega_\alpha = s_\alpha^* \omega$ on each open set U_α. Figure (3.1) gives an intuitive picture of the setup: a tangent vector T at x is lifted via the connection ω to a vector at a point on the curve which is the image under ϕ_α of the locus of $e \in G$, and this horizontal lift is then decomposed into two components, one along the fibre and the other tangent to the curve, with the former being defined to be $\omega_\alpha(T)$. The 'patching' condition is given in terms of the *canonical 1-form* θ on G, which is the unique left invariant $\mathfrak{g}$-valued 1-form determined by $\theta(A) = A$, $A \in \mathfrak{g}$. On the overlap $U_\alpha \cap U_\beta$, we set $\theta_{\alpha\beta} = \phi_{\alpha\beta}^* \theta$. Then we have on the overlap

$$\omega_\beta(T) = \mathrm{ad}((\phi_{\alpha\beta})_* T)^{-1})\,\omega_\alpha(T) + \theta_{\alpha\beta}(T). \tag{3.4.2}$$

It is not hard to translate (3.4.2) into the familiar formula (1.2.7) in gauge theory:

$$A_\mu^{(2)} = S A_\mu^{(1)} S^{-1} - ig dS\, S^{-1},$$

where the numerical factor ig is due just to a normalization convention introduced to relate mathematical entities to physical objects. Conversely, the connection form ω can readily be constructed from a family of 'local' 1-forms $\{\omega_\alpha\}$ satisfying (3.4.2), showing thus that the 'local' description is indeed equivalent to the preceding ones.

If we add to the definitions above the technical assumption that the manifold X is paracompact, then every principal bundle over X admits a connection. we have implicitly made such an assumption all through.

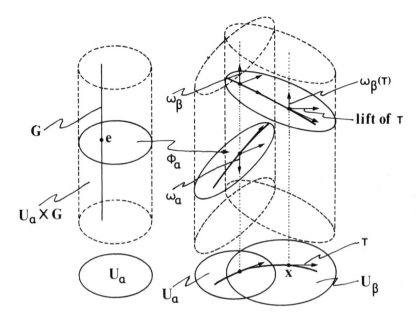

Figure 3.1: Defining local 1-forms ω_α

Covariant Derivative, Curvature and Holonomy

The idea of connection leads us to the concept of covariant differentiation. Suppose η is a p-form on P. Then we define its *covariant differential* or *exterior covariant derivative* by

$$D\eta \, (V_1, \dots, V_{p+1}) = (d\eta) \, (hV_1, \dots, hV_{p+1}), \tag{3.4.3}$$

where $d\eta$ is the ordinary differential and hV is the horizontal component of V. Again one can think 'locally' in the base space, since the left-hand side of (3.4.3) is independent of the vertical components. Given a vector field T on X, there is a unique vector field T^* on P which is horizontal and which projects onto T. We call T^* the *horizontal lift*, or just *lift*, of T. In each coordinate patch U_α, we can evaluate the 'local' form on tangent vectors $T_1, \dots, T_{p+1}$ to X by evaluating $d\eta$ on their lifts $T_1^*, \dots, T_{p+1}^*$. It is then easy to work out the patching condition.

Next, given a connection form ω on P, we define the corresponding *curvature form*

$$\Omega = D\omega. \tag{3.4.4}$$

Over each open set U_α, we have also the local curvature form:

$$\Omega_\alpha = D\omega_\alpha, \qquad (3.4.5)$$

satisfying the patching condition

$$\Omega_\beta(T) = \mathrm{ad}\left(((\phi_{\alpha\beta})_*T)^{-1}\right)\Omega_\alpha(T). \qquad (3.4.6)$$

The *structure equation* (of Cartan)

$$d\omega = -\omega\omega + \Omega \qquad (3.4.7)$$

can be rewritten in the familiar form

$$F_{\mu\nu} = \partial_\nu A_\mu - \partial_\mu A_\nu + ig[A_\mu, A_\nu], \qquad (3.4.8)$$

In this index-free notation, the *Bianchi identity* is simply

$$D\Omega = 0. \qquad (3.4.9)$$

Further, from the connection follows also the important concept of holonomy which was introduced before in gauge theory context as the phase factor (1.3.3) over closed loops. Consider a principal bundle $P \xrightarrow{\pi} X$. Let $\xi(t)$, $t = 0 \to t_1$, be a piecewise differentiable curve in X. Then a *horizontal lift*, or simply a *lift*, of $\xi(t)$ is a curve $\xi^*(t)$, $t = 0 \to t_1$, in P such that all its tangent vectors are horizontal. Through any point u such that $\pi(u) = \xi(0)$ there exists a unique lift of $\xi(t)$ which starts from u.

Suppose now we are given a curve $\xi(t)$ starting from x_0 and ending at x_1 in X. Let u_0 be an arbitrary point above x_0. Then the unique lift of $\xi(t)$ through u_0 ends in a point u_1 such that $\pi(u_1) = x_1$. By varying u_0 in the fibre above x_0 we get an isomorphism $\pi^{-1}(x_0) \xrightarrow{\sim} \pi^{-1}(x_1)$ which we call the *parallel transport* $\hat{\xi}$ of the fibre above x_0 to the fibre above x_1, along the curve $\xi(t)$. This is independent of the particular parametrization of the curve. In the case that the curve $\xi(t)$ is closed, i.e. $x_0 = x_1$, then $\hat{\xi}$ is a map of the fibre onto itself. By considering all piecewise differentiable closed curves through a point x_0, it is easy to see that the maps $\hat{\xi}$ form a group of automorphisms of the fibre, called the *holonomy group* $\Phi(u_0)$ through u_0, which can be identified with a subgroup of the structure group G. It can be shown that if X is connected, then all the holonomy groups through any $u \in P$ are conjugate to one another and are hence isomorphic. Therefore, if we are concerned only about the structure of the holonomy group given a connection and not which particular subgroup of G it is, we can omit the reference to the base point u_0. The *holonomy* of a given closed curve is just the parallel transport considered as an element of the structure group under the above embedding. The curvature at a given point can then be thought of as the holonomy of an infinitesimal closed curve through that point.

A connection of a principal bundle P is said to be *flat* if its curvature vanishes identically. If X is simply connected, then a flat connection implies that P is trivial. If X is connected but not necessarily simply connected, then the holonomy group Φ is discrete and is isomorphic to the quotient of $\pi_1(X)$ by a normal subgroup.

The theory of connection and holonomy in a principal bundle is much more intricate than we have presented here. There are times when the reference point of the holonomy matters. There are several subtly different concepts of holonomy groups. On the one hand we have the classical result of Ambrose and Singer that the Lie algebra of the holonomy group is determined by the curvature 2-form, and on the other we have the result of Wu and Yang that two gauge inequivalent potentials can give rise to the same field strength, even when the bundle is trivial over a contractible space (see Section 1.3). The physicist will find that he or she needs to deal with more and more refined concepts as such unexpected but fundamental facts arise.

Reduction of Bundles

Suppose $P \to X$, $Q \to X$ are two principal bundles with structure groups G and H respectively, where H is a subgroup of G. Then we say that Q is a *reduction* of P if there is a map $f : Q \to P$ satisfying, for R_a defined as in (3.4.1), $f(R_a u) = f(u)f(a)$, $u \in Q$, $a \in H$, such that the obvious maps induced from f are the embedding $H \to G$ and the identity map on X. We also say then that the bundle P is *reducible* to the subgroup H. This happens if and only if we can find open cover for X such that all the transition functions $\phi_{\alpha\beta}$ take their values in the subgroup H. Notice that P can be trivial without Q being so. This is the case of the t'Hooft–Polyakov monopole as discussed in Section 2.6. If in addition the bundles P and Q are given connections, then we say that the connection in P is *reducible* to the connection in Q if f maps horizontal spaces to horizontal spaces. In this case, the holonomy group of P is mapped onto the holonomy group of Q. Furthermore, the restriction of the connection form to Q is $\mathfrak{h}$-valued, where $\mathfrak{h}$ is the Lie algebra of H.

One result on the reduction of bundles which is of particular relevance to gauge theories uses the so-called *homotopy sequence* of bundles defined below. The derivation serves also as an explicit example of how exact sequences as introduced in Section 3.3 are applied. Given a fibre bundle

$$F \xrightarrow{i} E \xrightarrow{\pi} X$$

there is a long exact sequence called the *homotopy sequence* of the bundle as follows:

$$\cdots \to \pi_n(F) \xrightarrow{i_*} \pi_n(E) \xrightarrow{\pi_*} \pi_n(X) \xrightarrow{\partial} \pi_{n-1}(F) \to$$
$$\cdots \to \pi_1(X) \to \pi_0(F) \to \pi_0(E) \to \pi_0(X) \to 0.$$

The last four maps are somewhat complicated and we shall ignore them since they are not relevant to the subject matter of this book (see Bott and Tu (1982) and Steenrod (1951) for details).

Without actually going through the definition of the homomorphism ∂, let us apply this sequence in the case of the 't Hooft–Polyakov monopole (see Section 2.6). To simplify the arguments, let us first suppose the gauge group to be $SU(2)$. Then in the language of fibre bundles, we have a non-trivial $U(1)$ reduction of a trivial $SU(2)$-bundle. The non-triviality is defined by a homotopically nontrivial map

$$S^2 \to SU(2)/U(1)$$

which is essentially the value of the Higgs field at infinity. These topologically inequivalent field configurations are hence classified by:

$$\pi_2(SU(2)/U(1)).$$

Now by Example 2 above we have the principal bundle

$$U(1) \to SU(2) \to SU(2)/U(1) \cong S^2$$

and the relevant part of the homotopy sequence is

$$\begin{array}{ccccccc} \pi_2(SU(2)) & \to & \pi_2(SU(2)/U(1)) & \to & \pi_1(U(1)) & \to & \pi_1(SU(2)) \\ \| & & & & & & \| \\ 0 & & & & & & 0 \end{array}$$

By the general arguments given in Section 3.3 for exact sequences, this implies that the middle map in the above sequence is both injective and surjective or that it is an isomorphism:

$$\pi_2(SU(2)/U(1)) \cong \pi_1(U(1)) \cong \mathbb{Z}.$$

Hence we have the physically interesting result first discovered by 't Hooft and Polyakov that by choosing the appropriate Higgs field configuration, one can obtain all monopole charges.

If we now replace $SU(2)$ by $SO(3)$ with $\pi_1(SO(3)) = \mathbb{Z}_2$, as originally considered by 't Hooft and Polyakov, then we get the expected result that only even monopole charges are obtained, assuming we start with trivial $SO(3)$ monopole charge.

Characteristic Classes

Given any Riemannian manifold X, there are two bundles which are intrinsically and intimately related to its geometry. They are its tangent bundle and its frame bundle. The first is a vector bundle and the second a principal bundle. Conversely, it is often through the geometry of the base space that one finds the link between vector bundles and principal bundles.

Let $E \to X$ be an n-dimensional vector bundle, i.e. each fibre has dimension n. A *characteristic class* of dimension k is defined as a function $c(E) \in H^k(X)$ satisfying two conditions:

(i) $c(E) = c(E')$ if E and E' are isomorphic vector bundles, and

(ii) $c(f^*(E)) = f^*(c(E))$ for all continuous maps $f : X' \to X$,

where we recall that $f^*(E)$ denotes the pull-back of the bundle E. Here we have implicitly defined characteristic classes to take values in $\mathbb{R}$, but it can be shown that the ones we define here can in fact be normalized to take integer values. In the following, we shall include these normalization factors so that this is the case. The Stiefel-Whitney classes are defined modulo 2, but we shall not be concerned with them here.

Although characteristic classes are originally defined as invariants of vector bundles, it turns out that one can evaluate them using a connection in the associated principal bundle. Moreover, through the Weil homomorphism defined below, one can compute characteristic classes even for any arbitrary principal bundle.

For any Lie group G then with Lie algebra $\mathfrak{g}$, consider complex-valued functions $f(A_1, \ldots, A_k)$, $A_i \in \mathfrak{g}$, such that

(1) f is linear and totally symmetric in its arguments; and

(2) f is invariant, namely

$$f(a A_1 a^{-1}, \ldots, a A_k a^{-1}) = f(A_1, \ldots, A_k), \ \forall a \in G.$$

Because the function $f(A_1, \ldots, A_k)$ is linear in each A_i, it gives rise to a polynomial

$$f(A) = f(A, \ldots, A), \quad A \in \mathfrak{g},$$

of which $f(A_1, \ldots, A_k)$ is the complete polarization. For instance, if $f(A_1, A_2) = A_1 + A_1 A_2$ then the polynomial $f(A) = A + A^2$. We call $f(A)$ an *invariant polynomial*. All invariant polynomials under G form a ring $I(G)$.

The invariance condition (2) has the infinitesimal form:

$$\sum_{i=1}^{k} f(A_1,\dots,[B,A_i],\dots,A_k) = 0, \quad A_i, B \in \mathfrak{g}. \tag{3.4.10}$$

Given a principal bundle $P \xrightarrow{\pi} X$ with connection ω, it follows from (3.4.6) that if f is an invariant polynomial of degree k, then

$$f(\Omega) = f(\Omega_\alpha)$$

is a form of degree $2k$. The left-hand side shows that this form is globally defined in P, while the right-hand side shows that it is a form in X. By the Bianchi identity (3.4.9) and by (3.4.10) this form is closed and hence its cohomology class is an element of $H^{2k}(X)$. Furthermore, this class $[f(\Omega)]$ depends only on f and is independent of the choice of the connection ω. We have in fact defined the *Weil homomorphism*:

$$I(G) \longrightarrow H^*(X)$$
$$f \longmapsto [f(\Omega)].$$

The notation $H^*(X)$ is a short-hand way of writing $H^p(X)$ for the appropriate p.

For a complex vector bundle $E \to X$, the group $G = GL(n,\mathbb{C})$. The coefficients $f_k(A)$, $1 \le k \le n$, in the polynomial

$$\det(t + \frac{i}{2\pi}A) = t^n + f_1(A)t^{n-1} + \cdots + f_n(A), \tag{3.4.11}$$

where $A \in GL(n,\mathbb{C})$, are invariant polynomials. The corresponding cohomology classes

$$[f_k(\Omega)] = c_k(E) \in H^{2k}(X)$$

are the characteristic classes called the *Chern classes* of E.

It can be shown that any characteristic class of E is a polynomial in the Chern classes $c_1,\dots,c_n$. In particular the class

$$c(E) = 1 + c_1(E) + \cdots + c_n(E)$$

is called the *total Chern class* of E.

Each Chern class is represented by an invariant form, so that we can actually integrate them against suitable submanifolds of X. For concreteness, consider $U(2)$, and write

$$\Omega = \begin{pmatrix} ia & b \\ -\bar{b} & id \end{pmatrix}, \quad a, d \text{ real.}$$

We have

$$\det(t + \frac{i}{2\pi}\Omega) = t^2 - \frac{a+d}{2\pi}t + \frac{ad - b\bar{b}}{4\pi^4}.$$

Then

$$c_1 = \frac{1}{2\pi}\text{tr}\Omega$$

$$c_2 = \frac{1}{8\pi^2}(\text{tr}\Omega^2 - (\text{tr}\Omega)^2).$$

If the group is $SU(2)$, then $c_1 = 0$, and $c_2 = \frac{1}{8\pi^2}\text{tr}\Omega^2$.

For the $U(1)$ bundle over S^2 corresponding to monopole charge 1, consider the field strength given by

$$\Omega = \frac{1}{2r^3}(x^1 dx^2 \wedge dx^3 + x^2 dx^3 \wedge dx^1 + x^3 dx^1 \wedge dx^2).$$

Integrating the only non-trivial Chern class c_1 over S^2, we get

$$\int_{S^2} c_1 = \frac{1}{2\pi}\int_{S^2}\Omega = 1,$$

which is the magnetic charge in suitable units.

If instead of a complex vector bundle we consider a real bundle, very similar considerations apply and the corresponding characteristic classes

$$p_k(E) \in H^{4k}(X), \quad 1 \le k \le [\tfrac{m}{4}],$$

where $m = \dim X$, are called the *Pontrjagin classes* of E.

When applied to the tangent bundle of a manifold, the Pontrjagin classes are invariants of the differentiable structure. Similarly, the Chern classes of the tangent bundle of a complex manifold are invariants of the complex structure. Only the Stiefel–Whitney classes, which we did not study, are topological invariants of the underlying manifold.

Chapter 4

Loop Space Formulation

4.1 Why Loop Space, and Why Not?

In Section 1.3 we have seen already that, in contrast to the gauge potential $A_\mu(x)$, the phase factor $\Phi(C)$ in (1.3.3) is a physically measurable quantity, at least in theory, for any loop C in space-time. Further, in contrast to the field tensor $F_{\mu\nu}(x)$, $\Phi(C)$ can adequately describe all the physics contained in the gauge theory without any ambiguity. In principle, therefore, $\Phi(C)$ would be superior to both $A_\mu(x)$ and $F_{\mu\nu}(x)$ as dynamical variables.

In practice, however, loop variables are rather unwieldy for the following reason. $\Phi(C)$ in (1.3.3) is labelled by C, a closed curve in space-time, in contrast to $A_\mu(x)$ and $F_{\mu\nu}(x)$ both of which are labelled only by points x in space-time. Suppose we are working in 4-dimensional space-time; then each component of the fields $A_\mu(x)$ or $F_{\mu\nu}(x)$ represents just a 4-fold infinity of variables. In order to specify a closed curve C in space-time, on the other hand, we need to specify all the points on the curve C, so that a naive counting would suggest that the components of $\Phi(C)$ are labelled by an ∞-fold infinity of variables. Therefore, as dynamical variables for describing gauge theory, which was already adequately described by $\{A_\mu(x)\}$, $\{\Phi(C)\}$ must be highly redundant. To make effective use of $\{\Phi(C)\}$ as variables, this redundancy will have to be removed by imposing constraints on them, and the number of constraints required being infinite, the description in terms of $\{\Phi(C)\}$ is necessarily rather complicated.

Further, there is the question of how to represent a closed loop C. As will be seen in the next section, the most expedient way is to use parametrized loops, i.e. to regard each loop C as a map from the circle to space-time, namely a 4-vector function $\xi^\mu(s)$, $s = 0 \to 2\pi$, with $\xi(0) = \xi(2\pi)$. Then loop variables such as the phase factor $\Phi(C)$ are just functionals $\Phi[\xi]$ of the function $\xi = \{\xi^\mu(s);\ s = 0 \to 2\pi\}$. However, the actual geometric loop C in space-time is of course independent of the manner in which it is parametrized. Thus by considering parametrized loops we have further increased the variable redundancy which has also to be removed by imposing suitable extra constraints.

For these reasons, a so-called *loop space formulation* of gauge theory in terms entirely of these loop variables $\Phi(C)$, though aesthetically appealing, is seldom practically rewarding. Only in special situations where, for some reason, one has to work in any case with a large number of variables, or else one needs to keep explicit gauge invariance throughout, can one envisage that a loop space description will be profitable from the practical point of view. One area to which loop space techniques have been applied is the confinement problem in QCD, where the bonus is explicit gauge invariance. Another application to which we shall return is in the theory of monopoles where the necessity for patching will make the problem otherwise extremely complicated. Quite apart from practicability, however, once it is accepted that loop variables are what gauge theories really measure, the question of what constraints they have to satisfy so as to remove their redundancy is one of basic theoretical interest. The resolution of this question is the main concern of the present chapter.

4.2   Parametrized Loops and Derivative

To formulate gauge theory in loop space, our first task is to label the loop variables, or in other words to introduce coordinates for points in loop space. It will be seen that it is sufficient to consider only those loops passing through some fixed reference point, say $P_0 = \{\xi_0^\mu\}$ in space-time X. Such a loop can be parametrized as follows:

$$C : \{\xi^\mu(s);\ s = 0 \to 2\pi,\ \xi^\mu(0) = \xi^\mu(2\pi) = \xi_0^\mu\}, \qquad (4.2.1)$$

where $\xi^\mu(s)$ represent the coordinates in X of the points on the loop. The space of all such functions ξ^μ is the *parametrized loop space* introduced already in Section 3.1 as ΩX, whose 'points' are *parametrized loops*. The loop variables $\Phi(C)$ can then be written as the functional:

$$\Phi[\xi] = P_s \exp ig \int_0^{2\pi} ds\, A_\mu(\xi(s)) \frac{d\xi^\mu(s)}{ds}, \qquad (4.2.2)$$

where P_s denotes ordering in s increasing in our convention from right to left, and the derivative $d\xi^\mu(s)/ds$ in s, for a reason which will be apparent later, is to be taken from below.

Such a labelling of the loop variables Φ by a point in ΩX, as mentioned already in the preceeding section, is over-complete since the expression $\Phi[\xi]$ in (4.2.2) depends only on the loop C in X, not on the manner in which it is parametrized. By this we mean that if we introduce instead of s a new parameter, s' say, where $s' = f(s)$, it will only give a change in the variable of integration in (4.2.2) but will not change its value. Hence, if we define a Φ for each parametrization $\xi^\mu(s)$ of C, we have introduced a further redundancy into our loop variables. An alternative is not to distinguish the different parametrizations, but to label the loops C by the equivalence classes of functions $\xi^\mu(s)$ equivalent under reparametrizations. However, there seems to be no easy way to define derivatives and integrals in this quotient space of equivalence classes.

The reason to prefer parametrized loops is that ΩX has the virtue of being just a function space with familiar properties. For example, we can define loop differentiation in ΩX just as ordinary functional differentiation. Thus for any functional $\Psi[\xi]$ of ξ in ΩX:

$$\frac{\delta}{\delta \xi^\alpha(s)} \Psi[\xi] = \lim_{\Delta \longrightarrow 0} \frac{1}{\Delta} \{\Psi[\xi'] - \Psi[\xi]\}, \qquad (4.2.3)$$

with

$$\xi'^\mu(s) = \xi^\mu(s) + \Delta \delta_\alpha^\mu \delta(s - \bar{s}), \qquad (4.2.4)$$

where, to avoid ambiguity, $\delta(s - \bar{s})$ may be regarded as the zero width limit of a smooth function. As usual, the derivative so defined gives symmetric second derivatives when repeated:

$$\frac{\delta^2}{\delta \xi^\beta(s_2) \, \delta \xi^\alpha(s_1)} = \frac{\delta^2}{\delta \xi^\alpha(s_1) \, \delta \xi^\beta(s_2)}, \qquad (4.2.5)$$

property important for the development of differential and integral calculus akin to the usual one based on functions of points (see Section 3.3). On the other hand, derivatives for unparametrized loops are not so easily defined. To illustrate this point, we mention an example in the literature in which the loop derivative for unparametrized loops is defined as the continuum limit of the derivative on a discretized loop, as depicted in Figure 4.1a. The derivative defined in this way will in general give asymmetric second derivatives. The reason for this difference with the preceding case is quite transparent. Varying C at $\bar{s}$ first in direction α and then in direction β leads to Figure 4.1c, which differs from the result obtained from a variation in the opposite order, namely first β then α, giving Figure 4.1d. On the other hand, for the definition in terms of functional derivatives, the variation in either order results in the same loop parametrized by:

$$\xi''^\mu(s) = \xi^\mu(s) + \Delta \delta_\alpha^\mu \, \delta(s - \bar{s}) + \Delta' \delta_\beta^\mu \, \delta(s - \bar{s}), \qquad (4.2.6)$$

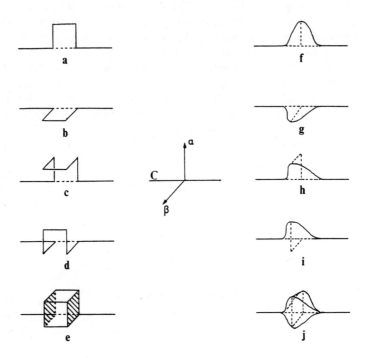

Figure 4.1: Illustrations for two versions of the loop derivative.

as illustrated in Figure 4.1h and 4.1i, which fact is necessary for obtaining symmetric mixed second derivatives.

From (4.2.3), we can define in particular the derivative in ΩX of the phase factor Φ. However, Φ being by definition an element of the gauge group for which the operation of addition is, strictly speaking, not defined, only its logarithmic derivative:

$$F_\mu[\xi|s] = \frac{i}{g}\,\Phi^{-1}[\xi]\,\frac{\delta}{\delta\xi^\mu(s)}\,\Phi[\xi], \qquad (4.2.7)$$

which is an element of the gauge Lie algebra, is meaningful. This quantity is of importance for our future discussions. From Figure 4.2, it is readily seen that in terms of ordinary field variables, we can write:

$$F_\mu[\xi|s] = \Phi_\xi^{-1}(s,0)\,F_{\mu\nu}(\xi(s))\,\Phi_\xi(s,0)\,\frac{d\xi^\nu(s)}{ds}, \qquad (4.2.8)$$

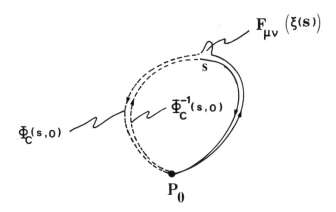

Figure 4.2: Illustration for the formula for $F_\mu[\xi|s]$.

where $F_{\mu\nu}(x)$ is the usual field tensor, and

$$\Phi_\xi(s_2, s_1) = P_s \exp ig \int_{s_1}^{s_2} ds\, A_\mu(\xi(s)) \frac{d\xi^\mu(s)}{ds} \qquad (4.2.9)$$

is the parallel transport in X from the point $\xi(s_1)$ to the point $\xi(s_2)$ along the loop C parametrized by ξ. In deriving (4.2.8) one may interpret (4.2.7) as the parallel phase transport first forwards to $\xi(s)$ along C followed by a detour at s, and then backwards again along the same loop. The phase factors for the segment of C beyond s cancel, but the factors for the remainder do not because of the detour at s giving thus the truncated phase factors $\Phi_\xi(s,0)$ and $\Phi_\xi^{-1}(s,0)$ in (4.2.8), while transport around the infinitesimal circuit occasioned by the detour at s is given by the field tensor $F_{\mu\nu}$ at that point $\xi(s)$.

One notes from (4.2.8) that $F_\mu[\xi|s]$ depends only on the 'early' part of the loop, i.e. it depends on $\xi^\mu(s')$ only for $s' \leq s$, not for $s' > s$, (see the remark on the derivative $d\xi/ds$ after (4.2.2)). This fact is reflected in the notation $[\xi|s]$ for its argument, which is henceforth to be so understood. Further, from the antisymmetry of $F_{\mu\nu}$ in its indices, it follows from (4.2.8) that $F_\mu[\xi|s]$ has components only transverse to the loop, namely that:

$$F_\mu[\xi|s] \frac{d\xi^\mu(s)}{ds} = 0. \qquad (4.2.10)$$

This statement is equivalent to the assertion that $\Phi[\xi]$ in (4.2.2) is invariant under reparametrization of the loop, since reparametrization means just shifting points along the loop, and so reparametrization invariance must therefore lead to a vanishing derivative along the loop direction. These properties of $F_\mu[\xi|s]$ will be of significance later when we consider the redundancy problem of loop variables.

4.3   Connection, Curvature and Holonomy

Geometrically, it is convenient to regard $F_\mu[\xi|s]$ as a gauge potential or, in the language of Section 3.4, as a connection in parametrized loop space ΩX since it represents the change in phase of Φ as one moves from a point ξ in ΩX to a neighbouring point $\xi + \delta\xi$, in the same way that the ordinary gauge potential $A_\mu(x)$ represents the parallel change in phase of a wave function ψ in moving from a point x to a neighbouring point $x + \delta x$ in space-time X. Now, given the potential (connection) $A_\mu(x)$ in X, one can construct from it the field tensor (curvature) $F_{\mu\nu}(x)$ via (1.2.12), and the phase factor (holonomy) $\Phi(C)$ via (1.3.3). So similarly, given now the 'connection' $F_\mu[\xi|s]$, one can construct also the corresponding 'curvature' and 'holonomy' in ΩX.

Thus, in parallel to (1.2.12), the 'loop curvature' is:

$$G_{\mu\nu}[\xi; s, s'] = \frac{\delta}{\delta\xi^\nu(s')} F_\mu[\xi|s] - \frac{\delta}{\delta\xi^\mu(s)} F_\nu[\xi|s'] + ig[F_\mu[\xi|s], F_\nu[\xi|s']]. \quad (4.3.1)$$

As usual for the curvature, $-igG_{\mu\nu}[\xi; s, s']\delta\xi^\mu(s)\delta\xi^\nu(s')$ represents the total change in phase as a point moves around an infinitesimal closed circuit. Here the circuit in ΩX is obtained starting from the point ξ by first varying ξ at s in the direction μ, then at s' in direction ν, then back again at s in direction μ and at s' in direction ν in the specified order. In particular, for $s = s'$, this circuit in ΩX will appear in ordinary space-time X like a skipping rope, sweeping out an infinitesimal 2-dimensional surface and enveloping a 3-dimensional element of volume, as illustrated in Figure 4.3b. On the other hand, for $s \neq s'$, no such volume is enclosed as depicted in Figure 4.3a. Hence, we expect that $G_{\mu\nu}[\xi; s, s']$ is zero for $s \neq s'$, i.e. it is proportional to $\delta(s - s')$. What its value is for $s = s'$ will depend on what lies inside the enclosed volume, which subject will be dealt with in the next section.

Just as for the connection $F_\mu[\xi|s]$, one can express the curvature $G_{\mu\nu}[\xi; s, s']$ in ΩX in terms of the usual field variables. This operation is a little delicate, and since the formula will be of particular use to us later, we shall derive it in some detail. We note first that by definition:

$$\frac{\delta}{\delta\xi^\kappa(s_2)} F_\lambda[\xi|s_1] = \lim_{\Delta \longrightarrow 0} \lim_{\Delta' \longrightarrow 0} \frac{1}{\Delta\Delta'} \frac{i}{g} \{\Phi^{-1}[\xi_2]\Phi[\xi_3] - \Phi^{-1}[\xi]\Phi[\xi_1]\}, \quad (4.3.2)$$

where for ξ given as in (4.2.1), we obtain:

$$\xi_1^\mu(s) = \xi^\mu(s) + \delta\xi^\mu(s), \tag{4.3.3}$$
$$\xi_2^\mu(s) = \xi^\mu(s) + \delta'\xi^\mu(s), \tag{4.3.4}$$
$$\xi_3^\mu(s) = \xi_1^\mu(s) + \delta'\xi^\mu(s), \tag{4.3.5}$$

through the variations:

$$\delta\xi^\mu(s) = \Delta\delta_\lambda^\mu \delta(s - s_1), \tag{4.3.6}$$
$$\delta'\xi^\mu(s) = \Delta'\delta_\kappa^\mu \delta(s - s_2). \tag{4.3.7}$$

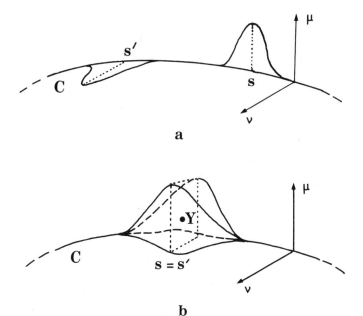

Figure 4.3: An infinitesimal closed circuit in ΩX seen in space-time.

By arguments similar to the derivation of (4.2.8), we may write:

$$\Phi[\xi_1] = \Phi[\xi] - ig \int ds \, \Phi_\xi(2\pi, s) \, F_{\mu\nu}(\xi(s)) \, \frac{d\xi^\nu(s)}{ds} \, \delta\xi^\mu(s) \, \Phi_\xi(s, 0), \tag{4.3.8}$$

and an analogous formula for $\Phi[\xi_2]$. For $\Phi[\xi_3]$, however, we have:

$$\Phi[\xi_3] = \Phi[\xi_1] - ig \int ds \, \Phi_{\xi_1}(2\pi, s) \, F_{\mu\nu}(\xi_1(s)) \frac{d\xi_1^\nu(s)}{ds} \, \delta'\xi_1^\mu(s) \, \Phi_{\xi_1}(s, 0). \quad (4.3.9)$$

Now in (4.3.9):

$$\begin{aligned}
\Phi_{\xi_1}(s, 0) &= \Phi_\xi(s, 0) - ig \int_0^s ds' \, \Phi_\xi(s, s') F_{\mu\nu}(\xi(s')) \frac{d\xi^\nu(s')}{ds'} \delta\xi^\mu(s') \Phi_\xi(s', 0) \\
&\quad + igA_\mu(\xi(s))\Phi_\xi(s, 0)\delta\xi^\mu(s),
\end{aligned} \quad (4.3.10)$$

where the last term in (4.3.10) is due to the variation of the end-point in the integral (4.2.9) for $\Phi_\xi(s, 0)$. A similar formula holds for $\Phi_\xi(2\pi, s)$. Furthermore, we have:

$$F_{\mu\nu}(\xi_1(s)) = F_{\mu\nu}(\xi(s)) + \partial_\rho F_{\mu\nu}(\xi(s)) \, \delta\xi^\rho(s), \quad (4.3.11)$$

and

$$\frac{d\xi_1^\nu(s)}{ds} = \frac{d\xi^\nu(s)}{ds} + \frac{d}{ds}(\delta\xi^\nu(s)). \quad (4.3.12)$$

Hence, collecting all variations to first order in $\delta\xi^\mu(s)$ and $\delta'\xi^\mu(s)$, we obtain:

$$\begin{aligned}
\frac{\delta}{\delta\xi^\kappa(s_2)} F_\lambda[\xi|s_1] &= ig[F_\kappa[\xi|s_2], F_\lambda[\xi|s_1]] \, \theta(s_1 - s_2) \\
&\quad + \Phi_\xi^{-1}(s_1, 0) D_\lambda F_{\kappa\nu}(\xi(s_1)) \Phi_\xi(s_1, 0) \frac{d\xi^\nu(s_1)}{ds_1} \delta(s_1 - s_2) \\
&\quad + \Phi_\xi^{-1}(s_2, 0) F_{\kappa\lambda}(\xi(s_2)) \Phi_\xi(s_2, 0) \frac{d}{ds_1} \delta(s_1 - s_2), \quad (4.3.13)
\end{aligned}$$

where D_λ denotes the usual covariant derivative defined in (2.2.9). Substituting (4.3.13) into (4.3.1) then gives:

$$\begin{aligned}
G_{\mu\nu}[\xi; s, s'] &= \Phi_\xi^{-1}(s, 0) \{D_\mu F_{\nu\rho}(\xi(s)) + D_\nu F_{\rho\mu}(\xi(s)) \\
&\quad + D_\rho F_{\mu\nu}(\xi(s))\} \, \Phi_\xi(s, 0) \frac{d\xi^\rho(s)}{ds} \delta(s - s'), \quad (4.3.14)
\end{aligned}$$

or equivalently in terms of the dual field tensor $^*F_{\mu\nu}$ defined in (2.2.8),

$$G_{\mu\nu}[\xi; s, s'] = \Phi_\xi^{-1}(s, 0)\epsilon_{\mu\nu\rho\sigma} D_\alpha \, ^*F^{\rho\alpha}(\xi(s))\Phi_\xi(s, 0) \frac{d\xi^\sigma(s)}{ds} \delta(s - s'), \quad (4.3.15)$$

which is the formula we sought for $G_{\mu\nu}$ in terms of ordinary field variables. One notes in particular that $G_{\mu\nu}[\xi; s, s']$ is indeed proportional to $\delta(s - s')$ as we have anticipated.

As explained in Sections 1.3 and 3.4, curvature is just the differential version of the global concept of holonomy (or phase factor). To construct the holonomy corresponding to the connection $F_\mu[\xi|s]$, let us parametrize a loop in loop space ΩX as follows:

$$\Sigma : \{\xi_t^\mu(s), \ s = 0 \to 2\pi, \ t = 0 \to 2\pi\}, \tag{4.3.16}$$

with

$$\xi_t^\mu(0) = \xi_t^\mu(2\pi) = \xi_0^\mu, \ t = 0 \to 2\pi, \tag{4.3.17}$$

$$\xi_0^\mu(s) = \xi_{2\pi}^\mu(s) = \xi_0^\mu, \ s = 0 \to 2\pi. \tag{4.3.18}$$

At each t, $\xi_t^\mu(s)$ represents for $s = 0 \to 2\pi$ a closed loop C_t in X passing through the reference point ξ_0:

$$C_t : \{\xi_t^\mu(s); s = 0 \to 2\pi\}. \tag{4.3.19}$$

For $t = 0$ or 2π, acording to (4.3.18), C_t shrinks to the point ξ_0. As t varies therefore, C_t traces out a closed loop in ΩX beginning and ending at the constant zero loop at ξ_0. The formula (4.3.16) thus represents a *parametrized loop* in ΩX, for which we can construct the holonomy or phase factor in complete analogy with (4.2.2) in X as:

$$\Theta(\Sigma) = P_t \exp ig \int_0^{2\pi} dt \int_0^{2\pi} ds \, F_\mu[\xi_t|s] \frac{\partial \xi_t^\mu(s)}{\partial t}, \tag{4.3.20}$$

with $F_\mu[\xi|s]$, the connection in ΩX, taking the place of the connection $A_\mu(x)$ in ordinary space-time X. The only difference is that loop space ΩX being infinite-dimensional, the connection $F_\mu[\xi|s]$ has many more components than $A_\mu(x)$, being labelled by μ as well as by the continuous index s, both of which have to be summed over in (4.3.20) — hence the additional integral over s compared with (4.2.2). Again P_t denotes ordering in the loop parameter t, with t increasing by convention from right to left, and $\partial \xi_t^\mu(s)/\partial t$ is taken to be the derivative from below.

In ordinary space-time X, the parametrized loop (4.3.16) in ΩX can be considered as a parametrized 2-dimensional surface swept out by the loops C_t and enclosing a 3-dimensional volume in X. As for $G_{\mu\nu}[\xi; s, s']$, the value of $\Theta(\Sigma)$ will depend on what is contained inside the volume enveloped by $\{\xi_t(s)\}$, a question which we shall now examine.

4.4 Monopoles as Sources of Curvature

Consider first the abelian theory. In that case, the covariant derivative D_μ is the same as the ordinary derivative ∂_μ and the phase factor $\Phi_\xi(s, 0)$

commutes with the field tensor $f_{\mu\nu}$, so that the formula (4.3.15) for the loop space curvature reduces simply to:

$$G_{\mu\nu}[\xi; s, s'] = \epsilon_{\mu\nu\rho\sigma} \partial_\alpha {}^* f^{\rho\alpha}(\xi(s)) \frac{d\xi^\sigma(s)}{ds} \delta(s - s'). \qquad (4.4.1)$$

Now at any point along a loop ξ at a value s, say, of the loop parameter, where $f_{\mu\nu}(\xi(s))$ is a gauge field in the sense that it is derivable from a gauge potential a_μ, $f_{\mu\nu}(\xi(s))$ must satisfy the Bianchi identity (2.2.3) which says that $\partial_\alpha {}^* f^{\rho\alpha}(\xi(s))$ vanishes, and hence also that $G_{\mu\nu}[\xi; s, s']$ vanishes. However, if a loop passes through a monopole, then at the value of s where the loop intersects the monopole world-line $Y(\tau)$, namely when $\xi(s) = Y(\tau)$ for some τ, then the above conclusion need no longer hold, since at the position of a monopole a_μ need not exist. Indeed, one has already seen in (2.2.5) that ${}^* f_{0i}$ is just the magnetic field $\mathbf{H}_i$, so that the 0th component of $\partial_\alpha {}^* f^{\rho\alpha}$ is just -4π times the magnetic charge, a fact which may also be written as:

$$\partial_\alpha {}^* f^{0\alpha}(x) = -4\pi\tilde{e} \int d\tau \, \frac{dY^0(\tau)}{d\tau} \, \delta(x - Y(\tau)). \qquad (4.4.2)$$

From Lorentz covariance, it then follows that:

$$\partial_\nu {}^* f^{\mu\nu}(x) = -4\pi\tilde{e} \int d\tau \, \frac{dY^\mu(\tau)}{d\tau} \, \delta(x - Y(\tau)), \qquad (4.4.3)$$

where the right-hand side represents the *magnetic current* carried by the monopole $\tilde{e}$. Substitution of (4.4.3) into (4.4.1) thus shows that, in this case at least, monopoles may be regarded as sources of loop space curvature.

Consider next a nonabelian theory. The assertion still holds that wherever $F_{\mu\nu}$ is a gauge field derivable from some potential A_μ, then the loop space curvature will vanish. As before also, at the point where a loop intersects the world-line of a monopole, the argument fails since at the position of a monopole, A_μ does not exist, so that the Bianchi identity need no longer be valid. However, it is not immediately clear from the formula (4.3.15) what the value of the loop space curvature should be at the monopole position, since the field tensor $F_{\mu\nu}$ and the covariant derivative D_μ (which depends on A_μ in this case) are now both patched quantities and are hence undefined there. To evaluate $G_{\mu\nu}[\xi; s, s']$ in that case, we shall have to analyse the problem further.

Initially, to avoid unnecessary confusion about infinitesimals, let us, as it were, examine the region around a monopole under a microscope, and consider, instead of the curvature, the holonomy over a finite-sized loop in ΩX. As explained in the last section, a parametrized loop in ΩX such as (4.3.16) represents a parametrized 2-dimensional surface Σ in ordinary space-time X, and our question now is what value the holonomy Θ over this loop in ΩX will take if the surface Σ it represents happens to enclose in ordinary space a monopole. This is exactly the situation depicted in Figure 2.1.

To be specific, consider first the simple example of the pure $\mathfrak{su}(2)$ Yang–Mills theory with gauge group $SO(3)$. Remembering the definition (4.2.7) of $F_\mu[\xi|s]$ as the logarithmic derivative of the phase factor $\Phi[\xi]$, one may write:

$$\exp ig \, dt \int_0^{2\pi} ds \, F_\mu[\xi_t|s] \, (\partial \xi_t^\mu(s)/\partial t) \approx \Phi^{-1}[\xi_{t+dt}] \, \Phi[\xi_t] \qquad (4.4.4)$$

in (4.3.20). Then $\Theta(\Sigma)$, being just a product ordered in t of such factors, is the total change in $\Phi[\xi_t]$ for t varying over the range $t = 0 \to 2\pi$. Now both Φ and Θ may be interpreted either as elements of the gauge group $SO(3)$ or as elements of the covering group $SU(2)$. Learning from our earlier experience in Chapter 2, we choose to regard them here as elements of $SU(2)$. In that case, if Σ encloses a monopole, then $\Phi[\xi_t]$ will trace out for $t = 0 \to 2\pi$ a curve in $SU(2)$ which winds only half way around the group in the manner explained in the paragraph before (2.5.1). Hence, in the notation used there, $\Theta(\Sigma)$, which measures the *total change*, is just $(\Phi^{(S)}[\xi_t])^{-1}\Phi^{(N)}[\xi_t]$ evaluated at $t = t_e$, giving it the value $-I$. On the other hand, if Σ does not enclose a monopole, then $\Phi[\xi_t]$ traces out a closed curve in $SU(2)$, giving $\Theta(\Sigma)$ the value I. In other words, denoting the monopole charge in an $SO(3)$ theory by $\zeta = \pm 1$, we may then write:

$$\Theta(\Sigma) = \zeta_\Sigma I, \qquad (4.4.5)$$

where ζ_Σ is the monopole charge enclosed in ordinary space by the surface Σ which corresponds to the loop in loop space ΩX parametrized by $\{\xi_t; t = 0 \to 2\pi\}$.

The formula (4.4.5) is interesting in that it gives us an explicit formula for evaluating the monopole charge in a nonabelian theory which has so far been defined only abstractly as a homotopy class. In a sense, the formula is just a direct extension of an already familiar concept in abelian theory. It was pointed out in (2.1.1) that the winding number n which labels the abelian monopole charge is given by $1/(2\pi)$ times the total change in phase of the factor $\Phi(C_t)$ as t goes from 0 to 2π. Now the phase of $\Phi(C_t)$ may be regarded as an element of the real line $\mathbb{R}$ which covers the gauge group $U(1)$, whose image in $U(1)$ via exponentiation is $\Phi(C_t)$. Thus, one sees that already in the abelian theory, the monopole charge may be interpreted as the total change, for t varying in the range $0 \to 2\pi$, of an element in the covering group corresponding to $\Phi(C_t)$ in the gauge group. Our result (4.4.5) above then appears just as a generalization in replacing the gauge group $U(1)$ by $SO(3)$ and the covering group $\mathbb{R}$ by

$SU(2)$. Further, these considerations show how similar results may also be derived in theories with other gauge groups. Thus, it can readily be shown that the result (4.4.5) in fact holds for all pure $\mathfrak{su}(N)$ Yang–Mills theories with gauge groups $SU(N)/\mathbb{Z}_N$, except that now the monopole charge can take values $\zeta = \exp i 2\pi r/N$, $r = 0, 1, 2, ..., (N - 1)$. With slight modifications, the results can also be extended to gauge groups $U(N)$.

Having now evaluated $\Theta(\Sigma)$ for a finite loop ξ in ΩX, we return to the question of the curvature $G_{\mu\nu}[\xi; s, s']$, the latter being, crudely speaking, the logarithm of Θ for an infinitesimal loop. For $s \neq s'$, $G_{\mu\nu}[\xi; s, s']$ corresponds to a Σ enclosing zero volume so that $\Theta(\Sigma) = I$, whose logarithm is by definition zero. The same value is obtained also for $s = s'$ if $\xi(s)$ does not sit on a monopole world-line, since though Σ now encloses a non-zero volume, it contains no monopole and $\Theta(\Sigma)$ again takes the value I. However, when $\xi(s)$ for $s = s'$ coincides with a monopole, $G_{\mu\nu}[\xi; s, s']$ corresponds to a Σ enclosing a monopole charge and takes on, according to (4.4.5), a value equal to the logarithm of the monopole charge at $\xi(s)$. Taken altogether then, these observations yield the following expression for the curvature:

$$
G_{\mu\nu}[\xi; s, s'] = -\frac{\pi}{g} \int d\tau \; \kappa[\xi|s] \epsilon_{\mu\nu\rho\sigma} \frac{d\xi^\rho(s)}{ds} \frac{dY^\sigma(\tau)}{d\tau} \, \delta^4(\xi(s) - Y(\tau)) \, \delta(s - s'),
$$

$$(4.4.6)$$

where $\kappa[\xi|s]$ satisfies:

$$
\exp i\pi\kappa = \zeta, \tag{4.4.7}
$$

ζ being the charge carried by the monopole moving along the world-line $Y(\tau)$. The equation has to be interpreted with some care, since, given ζ, the solution of (4.4.7) for κ is not unique in the gauge Lie algebra. We see, however, that when there is a monopole, say for $\zeta = -$ in the $SO(3)$ theory, $\kappa = 0$ is not a solution. We have thus demonstrated that, as in the abelian theory, monopoles in nonabelian theories also appear as sources of curvature in loop space.

4.5   Extended Poincaré Lemma

In the last section, we have spent considerable effort in clarifying the role that monopoles play in loop space. This is because monopoles will automatically enter in a very essential manner in any attempt to understand the problem of redundancy in loop variables which is central to the whole loop space formulation of gauge theories.

That this should be the case is perhaps not entirely unexpected. A parallel situation already exists in the abelian theory. Although classical electromagnetism can be described in terms of the antisymmetric field tensor $f_{\mu\nu}(x)$ as variables, these are not independent variables but must be expressible in terms of some potential via (1.1.6). If we think of $a_\mu(x)$ as the independent variables of the theory, then $f_{\mu\nu}(x)$ are redundant, having 6 components compared with only 4 for the gauge potential $a_\mu(x)$, and must therefore be constrained. The constraints necessary and sufficient for removing their redundancy are exactly the Bianchi identity (2.2.3). The fact that the two statements (1.1.6) and (2.2.3) are equivalent for antisymmetric $f_{\mu\nu}(x)$ is sometimes known as the Poincaré lemma (see Section 3.3). Now, the physical content of (2.2.3) is that there should be no monopole (or magnetic source) at x. Hence, we arrive at the conclusion that the absence of monopoles is the essence of the constraint required for removing the redundancy of $f_{\mu\nu}(x)$ as field variables.

Notice that the statements in the preceding paragraph are local in x and are not affected by what happens elsewhere. Suppose, for example, that there is a monopole moving along the world-line $Y^\mu(\tau)$, say, then the Bianchi identity no longer holds everywhere but is replaced by (4.4.3). In that case, $a_\mu(x)$ no longer exists on $Y^\mu(\tau)$, but since (4.4.3) still implies that (2.2.3) is satisfied for all $x \neq Y(\tau)$, $f_{\mu\nu}(x)$ is still expressible in terms of some gauge potential $a_\mu(x)$ at all these points.

Let us now turn to the nonabelian theory. We are interested in the parallel problem of what constraints are to be imposed on loop variables so as to remove their redundancy. We shall work with the variables $F_\mu[\xi|s]$ which, as we saw in (4.2.8), are closely related to the field tensor $F_{\mu\nu}(x)$ and yet share with the abelian field tensor $f_{\mu\nu}(x)$ the important property of being gauge invariant, and not merely gauge covariant as $F_{\mu\nu}(x)$. To address the problem of redundancy for $F_\mu[\xi|s]$, let us first collect together the conditions required for removing the redundancy of $f_{\mu\nu}(x)$ in the abelian theory:

(a) that $f_{\mu\nu}(x)$ is a local quantity depending on the space-time point x,

(b) that $f_{\mu\nu}(x)$ is antisymmetric under interchange of the indices μ and ν,

(c) that $f_{\mu\nu}(x)$ should have no monopoles.

By analogy, we are tempted to suggest that perhaps for the general theory, including the nonabelian case, the constraints required for removing the redundancy in $F_\mu[\xi|s]$ would be three parallel conditions with the same physical content. Now from (4.2.10), we have already seen that the equivalent of (b) for the variable $F_\mu[\xi|s]$ is that it should have only components which are transverse to the loop C as parametrized by the function $\xi(s)$. Further, we saw in (4.2.8) that $F_\mu[\xi|s]$ depends only on $\xi(s')$ for $s' \leq s$, as symbolized in its argument $[\xi|s]$, and that although this property has to do with the path-ordering of operators in the nonabelian theory which has no strict parallel in the abelian, the arguments which led to it when reduced to the abelian case imply the statement (a) that $f_{\mu\nu}(x)$ is a local quantity. Hence, we suggest that so long as we remember the meaning of the notation $[\xi|s]$ and work with $F_\mu[\xi|s]$ having only transverse components, we shall be left with only the parallel of (c) to consider.

As far as we know, the statement that there is no monopole charge at a certain point x cannot be made for nonabelian theories in the form of a space-time local condition like the Bianchi identity of (2.2.3). However, as seen in the preceding section, it can be readily formulated in loop space, in terms of precisely the variables $F_\mu[\xi|s]$ we are using, either as (4.4.5) with $\zeta_\Sigma = 1$ for any infinitesimal Σ surrounding x, or else as (4.4.6) with $\kappa[\xi|s] = 0$ for all ξ and s satisfying $\xi(s) = x$. However, in order to remove completely the redundancy from our loop variables $F_\mu[\xi|s]$, we have to ensure that there are no monopoles not just at some points x but everywhere except at the positions of those monopoles already specified. This will be the case if and only if the condition (4.4.5) is satisfied for all surfaces Σ, or equivalently, if the condition (4.4.6) is satisfied for all points s on all parametrized loops ξ.

In analogy then with the usual Poincaré lemma for the abelian theory, we suggest an extended Poincaré lemma for the general nonabelian theory which states that the necessary and sufficient conditions for removing the redundancy from $F_\mu[\xi|s]$ are as follows:

(A) that $F_\mu[\xi|s]$ depends on $\xi(s')$ only for $s' \leq s$,

(B) that $F_\mu[\xi|s]$ has only components transverse to the loop parametrized by ξ, namely that it satisfies the condition (4.2.10) for all ξ and s,

(C) that $F_\mu[\xi|s]$ satisfies the condition (4.4.5) for all surfaces Σ, or equivalently the condition (4.4.6) for all ξ and s.

We recall that by the removal of redundancy from $F_\mu[\xi|s]$ we mean the guaranteed existence everywhere, except at the specified positions of the mono-poles, of a gauge potential $A_\mu(x)$ which is related to $F_\mu[\xi|s]$ via (4.2.7) and (4.2.2) for every ξ and s, since only then can we replace $A_\mu(x)$ by $F_\mu[\xi|s]$ as variables and yet be sure that we are still dealing with the same gauge theory.

That the extended Poincaré lemma suggested above indeed holds has been demonstrated in many ways with varying degrees of rigour. Nor is the above the only way of stating the lemma, but is the one we find most physically illuminating. We give below an outline of a 'proof', which though far from rig-orous, will, we hope, give sufficient confidence in the lemma's validity. Again, it would be enough to illustrate with the simple example of an $SO(3)$ theory.

That the conditions (A), (B), and (C) are necessary to remove the redun-dancy in the variables $F_\mu[\xi|s]$ is obvious, since these properties of $F_\mu[\xi|s]$ were originally deduced from the expressions (4.2.7) and (4.2.2). It is the converse statement that they are also sufficient for removing the redundancy that is more difficult to verify.

To do so, define first, in analogy with (4.2.9), the quantity:

$$\Theta_\Sigma(t_2, t_1) = P_t \exp ig \int_{t_1}^{t_2} dt \int_0^{2\pi} ds \, F_\mu[\xi_t|s] \frac{\partial \xi_t^\mu(s)}{\partial t} \qquad (4.5.1)$$

as the parallel transport in parametrized loop space ΩX from the 'point' ξ_{t_1} to the 'point' ξ_{t_2} along the 'path' Σ as parametrized by the 'function' ξ_t of t for $t = 0 \to 2\pi$. The condition (C) above which gives for $SO(3)$:

$$\Theta(\Sigma) = \pm 1, \qquad (4.5.2)$$

for all 'closed loop' Σ in ΩX then implies that $\Theta_\Sigma(t, 0)$, for fixed 'end-point' ξ_t, can depend on the 'path' Σ only through a sign. This means that $\Theta_\Sigma(t, 0)$ defines a unique element of the gauge group $SO(3)$ which we may call $\Phi^{-1}[\xi_t]$ being dependent only on the 'end-point' ξ_t but not otherwise on the 'path' Σ. Furthermore, the $\Phi[\xi]$ so defined satisfies (4.2.7) as desired, since by definition

$$\Phi^{-1}[\xi_t] \frac{\delta}{\delta \xi_t^\mu(s)} \Phi[\xi_t] = \lim_{\Delta \to 0} \left\{ P_{t'}^{-1} \exp -ig \int_t^{t+\Delta} dt' \int_0^{2\pi} ds' F_\alpha[\xi_{t'}|s'] \frac{\partial \xi_{t'}^\alpha(s')}{\partial t'} - 1 \right\},$$
$$(4.5.3)$$

where P^{-1} denotes reverse path-ordering and

$$\xi_{t+\Delta}^\alpha(s') = \xi_t^\alpha(s') + \Delta \delta_\mu^\alpha \delta(s - s') \qquad (4.5.4)$$

which gives exactly $-ig F_\mu[\xi_t|s]$ as expected.

That $\Phi[\xi|s]$ so defined is independent of parametrization, namely that it depends really only on the loop C that the function ξ represents rather than on the function ξ itself, is also obvious because of the condition (B) on $F_\mu[\xi|s]$, as explained in the passage after (4.2.10).

To show finally that $\Phi(C)$ so defined is indeed of the form (4.2.2), we need first to verify that it satisfies a composition law, which it obviously does if it is of form (4.2.2), namely that:

$$\Phi(C^2)\Phi(C^1) = \Phi(C^2 * C^1), \qquad (4.5.5)$$

where $C^2 * C^1$ is the loop obtained by first going round C^1 and then C^2, as illustrated in Figure 4.4. We can do so as follows. Let C^1 and C^2 be any two loops passing through the reference point. For each, we choose a parametrized surface, say $\{\xi_t^i(s); s, t = 0 \to 2\pi\}$, $i = 1, 2$, such that at some given value of t, say π, ξ_π^i represents just the loop C^i so that we can, according to the discussion

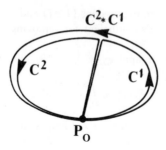

Figure 4.4: Composition of two loops.

above, equate $\Phi(C^i)$ to $\Theta_{\Sigma^i}^{-1}(\pi,0)$ as defined in (4.5.1). Next, we shall try to construct, by a suitable reparametrization of the surfaces Σ^i represented by ξ_t^i, a third parametrized surface, say, $\{\xi_t^*(s); s,t = 0 \to 2\pi\}$ passing through the loop $C^2 * C^1$ say again at $t = \pi$, such that:

$$\Theta_{\Sigma^*}(\pi,0) = \Theta_{\Sigma^1}(\pi,0)\,\Theta_{\Sigma^2}(\pi,0). \tag{4.5.6}$$

If we succeed in constructing such a parametrized surface $\{\xi_t^*\}$, then our arguments above will imply that $\Phi(C^2 * C^1)$, which is by definition $\Theta_{\Sigma^*}^{-1}(\pi,0)$, indeed satisfies the composition law (4.5.5) as required. The idea is that although such a construction is explicitly parametrization dependent, the result (4.5.5) will not be since, being a relation between parametrization independent quantities, it will acquire a general validity beyond that particular construction.

An example of such a construction is illustrated in Figure 4.5, where for $t = 0 \to \pi$, C_t^* remains at the fixed reference point P_0, while for $s = 0 \to \pi$, it sweeps over the surface Σ^2 arriving at $t = \pi/2$ at a reparametrization of the loop C^2. Then for $t = \pi/2 \to \pi, s = 0 \to \pi$, C_t^* sweeps over the surface Σ^1 arriving for $t = \pi$ at a reparametrization of the loop C^1, while for $s = \pi \to 2\pi$ it remains unchanged at C^2. Thus for $t = 0 \to \pi$, the loop C_t^* sweeps over a surface composed of Σ^1 and Σ^2 arriving for $t = \pi$ at $C_\pi^* = C^2 * C^1$ as we wanted. Further, one can see by a suitable reparametrization of the surfaces that (4.5.6) is satisfied by construction, where we have to use the condition (A) that $F_\mu[\xi|s]$ is independent of $\xi(s')$ for $s' > s$. As a result we have (4.5.5) as desired.

Having then both parametrization independence and the composition law for $\Phi(C)$, it is then straightforward, at least in the case when there are no monopoles, to construct a gauge potential $A_\mu(x)$ such that (4.2.2) is satisfied for any parametrization ξ of C. For each point x in space-time, we choose a

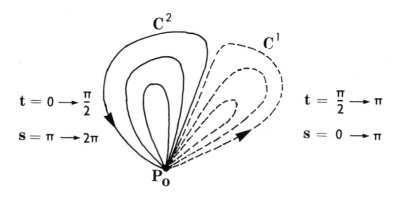

Figure 4.5: Illustration for proving the composition law for $\Phi(C)$.

path, γ_x say, joining the point x to the reference point P_0 in such a way that neighbouring points are joined to P_0 by neighbouring paths. The simplest would be to take γ_x just as the straight line joining x to P_0. Then, for any point x' close to x and $\gamma_{x'x}$ being the straight line joining x to x', the composition $\gamma_{x'}^{-1} * \gamma_{x'x} * \gamma_x$ is a closed loop, as illustrated in Figure 4.6, for which one can define a phase factor $\Phi(\gamma_{x'}^{-1} * \gamma_{x'x} * \gamma_x)$, which being dependent only on the points x and x' but not on the parametrization of the paths, we shall denote as $h(x', x)$. We define then the gauge potential as:

$$A_\mu(x) = -\frac{i}{g} \lim_{\Delta \to 0} \{h(x', x) - 1\} \qquad (4.5.7)$$

for

$$x'^\nu = x^\nu + \Delta \delta_\mu^\nu. \qquad (4.5.8)$$

The proof that $A_\mu(x)$ so defined does indeed satisfy (4.2.2) for any parametrized loop ξ can be illustrated as in Figure 4.7. By virtue of the composition law (4.5.5) and the definition of $h(x', x)$ above, one sees that $\Phi[\xi]$ for any ξ can be written as:

$$\Phi[\xi] = P_s \prod_s h\left(\xi(s) + \frac{d\xi(s)}{ds} ds, \; \xi(s)\right), \qquad (4.5.9)$$

where since for ds infinitesimal,

$$h\left(\xi(s) + \frac{d\xi(s)}{ds} ds, \; \xi(s)\right) = \exp ig\left\{A_\mu(\xi(s)) \frac{d\xi^\mu(s)}{ds} ds\right\}, \qquad (4.5.10)$$

we have (4.2.2) as desired.

In the situation when there are no monopoles anywhere in space-time, we have then shown that the conditions (A), (B), and (C) are indeed sufficient to guarantee the existence of a gauge potential $A_\mu(x)$, or in other words to remove the redundancy from the variables $F_\mu[\xi|s]$ as was suggested. The potential we have constructed depends on the choice of a path γ_x for each point x in space-time, which we have taken above to be the straight line joining the reference point P_0 to x. Such a choice seems quite arbitrary, and it would be surprising if it has any real physical significance. Indeed, had we chosen another set of paths γ'_x joining each point x to the reference point P_0, we would have obtained another potential $A'_\mu(x)$ derived from:

$$h'(x', x) = \Phi(\gamma'^{-1}_{x'} * \gamma_{x'x} * \gamma'_x) \tag{4.5.11}$$

in a manner similar to (4.5.7). However, by the composition law (4.5.5), we

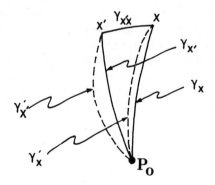

Figure 4.6: Construction of the gauge potential from loops.

may write:

$$\Phi(\gamma'^{-1}_{x'} * \gamma_{x'x} * \gamma'_x) = \Phi(\gamma'^{-1}_{x'} * \gamma_{x'})\Phi(\gamma^{-1}_{x'} * \gamma_{x'x} * \gamma_x)\Phi(\gamma^{-1}_x * \gamma'_x), \tag{4.5.12}$$

as illustrated in Figure 4.6. Then by defining

$$S(x) = \Phi(\gamma^{-1}_x * \gamma'_x), \tag{4.5.13}$$

we have:

$$h'(x', x) = S(x')h(x', x)S^{-1}(x), \tag{4.5.14}$$

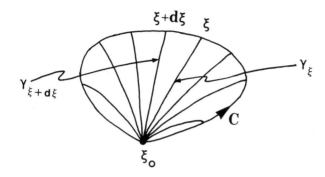

Figure 4.7: Construction of loops from the gauge potential.

or, equivalently by (4.5.7), that $A'_\mu(x)$ is related to $A_\mu(x)$ by a gauge transformation parametrized by $S(x)$ in (4.5.13). In other words, we have shown that a change in the choice of γ_x is just a change in gauge for $A_\mu(x)$, or that the construction above for the gauge potential is unique up to gauge transformations.

What happens in the case when there are monopoles around? The situation is then a little more complicated. We recall that in order for the above construction to work, we need to assign a path γ_x joining the reference point P_0 to each point x in the domain D for which the gauge potential $A_\mu(x)$ is meant to exist in such a way that neighbouring points in D should correspond to neighbouring paths. When there are monopoles around, the domain D is no long the whole of space-time, but space-time minus the world-lines of the monopoles which is not a simply-connected domain. Our previous choice of γ_x as the straight lines joining P_0 to x are then not good enough since two straight lines joining P_0 to two neighbouring points may be separated by a monopole world-line and will not approach each other even when the two points do so. However, given the result in (4.5.14) above, we can still proceed with the construction so longer as we are able to specify some paths γ_x with this property. That one can indeed find such paths even when there are monopoles can be seen in the following example.

It is sufficient to illustrate with the simplest example of having only one monopole. At each time, we cover the the space around the monopole by two patches (N) and (S) as in (2.3.7). The domain D is then covered by two patches obtained by taking the union of (N) and (S) respectively over all times. Our task now is to specify a path γ for each point in (N) such that neighbouring points are joined to P_0 by neighbouring paths. The method above will then allow us to construct a gauge potential $A_\mu(x)$ for the (N) patch. Similarly, we can try the same with the southern patch. Finally, to complete the proof, we have to demonstrate that the potentials so constructed are related by the proper patching transformation. A possible choice of γ_x for the northern patch can be as follows. For a point x at the same time as the reference point P_0, we take γ_x to be the straight line joining P_0 to some fixed point, say P_N, lying on the northern polar axis together with the straight line joining P_N to x. For any point not at the same time as P_0, we extend the path further by another piece running parallel to the monopole world-line. We notice that by construction, the first part of the path from P_0 to P_N is the same for all points in (N), while the remaining part of the path lies entirely in (N). Furthermore, for two points x and x' which are neighbouring in (N), the straight line $\gamma_{x'x}$ joining them also lies entirely in (N). Hence, when the two points approach each other, their two paths γ_x and $\gamma_{x'}$ will do the same, satisfying thus the necessary criterion, and allowing us to construct the gauge potential $A_\mu^{(N)}(x)$ as we did above. We then repeat the construction of $A_\mu^{(S)}(x)$ in an analogous fashion for points in the southern patch. Finally, to show that the potentials $A_\mu^{(N)}(x)$ and $A_\mu^{(S)}(x)$ satisfy the patching condition, we pick a point in the overlap region. The potentials constructed as above for such a point will depend on whether we consider it as a point in the northern or southern patch since the choice of γ_x differs in the two patches. However, the result in (4.5.14) above ensures that the two values of the gauge potentials so obtained are related by a patching gauge transformation with the patching function:

$$S_{SN} = \Phi\left(\gamma_x^{(N)^{-1}} * \gamma_x^{(S)}\right). \qquad (4.5.15)$$

This then completes our verification that the potential exists also when monopoles are around. We note however that in this case, although the potential is still related to $\Phi(C)$ by (4.2.2) even for C which lies partly in (N) and partly in (S), in evaluating the integral (4.2.2) in the overlap region the proper patching transformations will have to be performed.

The importance of the extended Poincaré lemma which we have just 'proved' lies in the fact that given any potential $A_\mu(x)$ we have a unique set of loop space variables $F_\mu[\xi|s]$ satisfying the conditions (A), (B) and (C), and also conversely, given any set of $F_\mu[\xi|s]$ satisfying (A), (B) and (C), we have a gauge potential $A_\mu(x)$ which is unique up to gauge transformations. This means that, by imposing the constraints (A), (B) and (C), we can now legitimately employ the loop variables $F_\mu[\xi|s]$ for a description of gauge theories which is completely equivalent to the standard description in terms of the gauge potential but yet is explicitly gauge invariant, apart from a quite harmless

x-independent gauge rotation at the reference point P_0.

The demonstrated existence of a gauge potential A_μ for the above framework also justifies the claim made at the beginning that it is sufficient to consider only loops passing through the reference point P_0, since given now A_μ, one can construct the phase factor $\Phi(C)$ for any loop C in space-time.

Chapter 5

Dynamics

5.1 Interaction as Consequence of Topology

By dynamics, we mean the laws governing the evolution in time of physical systems. Usually, these are given in terms of an action principle. For gauge theories, because of the basic assumption of local gauge invariance, the action has to be an invariant under local gauge transformations. For example, for the electromagnetic field, the standard action is:

$$\mathcal{A}_F^0 = -\frac{1}{16\pi} \int d^4x\, f_{\mu\nu}(x) f^{\mu\nu}(x), \qquad (5.1.1)$$

which gives the correct equations of motion as deduced from experiment. It happens also to be the simplest $U(1)$-gauge invariant one can write down. Although one can in principle ask why this alone should be chosen by Nature as the action out of many possible gauge invariants, it is not easy to find a convincing answer.

It is interesting, however, to query conceptually the usual manner in which interactions between the gauge field and matter are introduced. Consider first the simplest case of a classical charged particle placed in an electromagnetic field. The interaction between the charge and the field is conventionally formulated as follows. First, we add to (5.1.1) the free action of the particle:

$$\mathcal{A}_M^0 = -m \int d\tau, \qquad (5.1.2)$$

where the integral is to be taken along the world-line of the particle, with τ being the proper time along the world-line. To the total free action of the two systems, we then add an *interaction term*:

$$\mathcal{A}_I = e \int a_\mu(x) dx^\mu, \qquad (5.1.3)$$

the integral being also taken along the particle world-line. If we now vary the total action with respect to the dynamical variables, namely the gauge potential $a_\mu(x)$ and the coordinates $Y^\mu(\tau)$ of the particle, we obtain as Euler–Lagrange equations respectively the Maxwell equation:

$$\partial_\nu f^{\mu\nu}(x) = -4\pi e \int d\tau \frac{dY^\mu(\tau)}{d\tau} \delta(x - Y(\tau)), \qquad (5.1.4)$$

and the Lorentz equation:

$$m\frac{d^2 Y^\mu(\tau)}{d\tau^2} = -e f^{\mu\nu}(Y(\tau))\frac{dY_\nu(\tau)}{d\tau}, \qquad (5.1.5)$$

each coupling the field variable $a_\mu(x)$ to the particle coordinate $Y^\mu(\tau)$. Hence, we have interactions between the charged particle and the electromagnetic field.

One sees that when approached in this way, interactions appear as something of an afterthought, introduced somewhat arbitrarily as an addendum to the free action. The particular choice of (5.1.3) as the interaction term was basically *phenomenological*, meaning that one already had an idea what the coupled equations of motion should be as deduced by Faraday and Maxwell from experiment and chose the action accordingly. On the other hand, interactions being of such basic importance in physics, one could have hoped that they would arise *naturally* and be in some sense unique.

There is indeed a way of approaching the above problem in such a manner that the interaction of the charge with the electromagnetic field appears *naturally* as a unique consequence of topology without introducing the concept of an interaction term into the action. The arguments go as follows. We note first that in (5.1.4),

$$\partial_\nu f^{\mu\nu}(x) = 0 \qquad (5.1.6)$$

everywhere except on the world-line $Y^\mu(\tau)$ of the charged particle. Remembering then that by the definition of the *-operation (2.2.4):

$$f_{\mu\nu}(x) = -{}^*({}^*f_{\mu\nu}(x)), \qquad (5.1.7)$$

we may interpret (5.1.6) as the Bianchi identity for the dual Maxwell field ${}^*f_{\mu\nu}$. Hence, we may conclude by the Poincaré lemma that, except on the world-line $Y^\mu(\tau)$, there must exist at least locally a potential, say $\bar{a}_\mu(x)$, such that:

$$ {}^*f_{\mu\nu}(x) = \partial_\nu \bar{a}_\mu(x) - \partial_\mu \bar{a}_\nu(x), \qquad (5.1.8)$$

or that ${}^*f_{\mu\nu}$ is also a gauge field. Now in Section 2.1, we have shown that a monopole of the Maxwell field $f_{\mu\nu}$ is a magnetic charge which bears the same relationship to the magnetic field as an electric charge does to the electric field. Hence, since the *-operation just interchanges the electric and magnetic fields, it follows that a monopole of the Maxwell field $f_{\mu\nu}$, can also be regarded a source of the dual Maxwell field $^*f_{\mu\nu}$ in the same sense that an electric charge is regarded as a source of the Maxwell field $f_{\mu\nu}$. The same arguments inverted can now be applied to the dual field $^*f_{\mu\nu}$, which is also a gauge field as shown by (5.1.8), to deduce that a monopole in $^*f_{\mu\nu}$ is equivalent to a source of the Maxwell field $f_{\mu\nu}$, i.e. an electric charge such as that carried by $Y^\mu(\tau)$ in (5.1.4).

However, a monopole has an intrinsic interaction with the gauge field which is dictated by its topology. That this is the case can be seen intuitively as follows. We have already made clear before that the occurrence of a monopole charge at a given point means that the gauge field has a particular topological configuration in the spatial region surrounding that point. Suppose now that the monopole is displaced to another position. Then, in order to keep the same monopole charge, which we know from Section 2.1 must be conserved by continuity, the gauge field will have to rearrange itself so as to keep the same topological configuration around the new point. This means therefore that the gauge field described by the potential a_μ must depend in some way, or in other words be coupled to, the coordinates Y^μ of the monopole. In physical language, then, we have an interaction between the monopole and the gauge field which is inherent already in the topological definition of the monopole charge.

Since we have already concluded above that an electric charge can also be considered as a monopole of the dual field $^*f_{\mu\nu}$, it follows that it has an intrinsic interaction with the field by virtue alone of its topological structure. The only question is what form this interaction will take and whether it will be the same as that deduced in the conventional formulation from the interaction term (5.1.3) in the action. Let us attempt to answer this question.

It will be seen that the arguments we shall use to derive the explicit form of the intrinsic monopole–field interaction suggested above are in fact the same whether we are dealing with a *magnetic* monopole of the Maxwell field $f_{\mu\nu}$, or with an *electric* monopole of the dual field $^*f_{\mu\nu}$. Let us therefore work it out for the *magnetic* monopole for easier reference to the material presented earlier in Chapter 2. The problem may be precisely formulated as follows. Given the free action of the field–plus–particle system:

$$\mathcal{A}^0 = \mathcal{A}^0_F + \mathcal{A}^0_M, \tag{5.1.9}$$

for $\mathcal{A}^0_F$ and $\mathcal{A}^0_M$ defined as in (5.1.1) and (5.1.2), we wish to find the Euler–Lagrange equations extremizing $\mathcal{A}^0$ under the constraint that the particle at

$Y^\mu(\tau)$ should carry a monopole (magnetic) charge $\tilde{e}$. The variables in the problem are the gauge potential $a_\mu(x)$ and the particle coordinates $Y^\mu(\tau)$, under variations of which the action is to be extremized. At first sight, the problem looks technically rather complicated if we remember that in the presence of a monopole, the gauge potential $a_\mu(x)$ has to be patched and that the patches themselves depend on the monopole coordinates $Y^\mu(\tau)$ so that if we vary $Y^\mu(\tau)$, the patches will have to be redefined. Indeed, for this reason, when the problem was first stated and solved by Wu and Yang in 1976 as part of their larger programme, the procedure they presented used an ingenious argument which avoided a direct solution of the problem. We shall see, however, that by a judicious change of variables, the problem can be reduced to a beautifully simple form.

To see this, let us first recall that in the abelian theory, the topological constraint defining a monopole charge $\tilde{e}$ was shown in Section 2.1 to be equivalent to the flux condition:

$$\iint f_{\mu\nu}\,d\sigma^{\mu\nu} = -4\pi\tilde{e}, \qquad (5.1.10)$$

where the integral is taken over any surface enclosing the monopole. Next, using Gauss' theorem, the left-hand side of (5.1.10) can in turn be written as a volume integral of a divergence. Now, since the equation is required to be valid for any volume containing the monopole, it has to hold differentially, giving then in covariant notation the equation:

$$\partial_\nu\,{}^*f^{\mu\nu}(x) = -4\pi\tilde{e}\int d\tau\,\frac{dY^\mu(\tau)}{d\tau}\,\delta^4(x - Y(\tau)), \qquad (5.1.11)$$

which is to be satisfied for all x and μ. Introducing for each x and μ a Lagrange multiplier, say $\lambda_\mu(x)$, we obtain the auxiliary action incorporating the constraint:

$$\mathcal{A}' = \mathcal{A}^0 + \int dx\,\lambda_\mu(x)\left\{\partial_\nu\,{}^*f^{\mu\nu}(x) + 4\pi\tilde{e}\int d\tau\,\frac{dY^\mu(\tau)}{d\tau}\,\delta^4(x - Y(\tau))\right\}. \quad (5.1.12)$$

Equations of motion are now to be derived by extremizing $\mathcal{A}'$ with respect to unconstrained variations in $a_\mu(x)$ and $Y^\mu(\tau)$. In the absence of the constraint term, extremization of $\mathcal{A}^0$ would yield of course just the equations of the free field and the free particle. With the constraint term, one will obtain instead equations coupling $a_\mu(x)$ to $Y^\mu(\tau)$, namely equations comprising interactions between the charged particle and the field.

In view of the fact that a_μ has to be patched, making the variational problem directly in terms of a_μ cumbersome, we now propose to use instead the field tensor $f_{\mu\nu}$ as field variables. Normally, this will not do. Although we know from the discussion in, for example, Section 1.3, that the present classical problem of a point charge can adequately be described by the field tensor $f_{\mu\nu}$ alone, this last quantity has six independent components, even after taking account of its antisymmetry under the interchange of μ and ν. This compares with only four for the original variables a_{μ} which are already sufficient for a full description of the system. The field tensor, as variables, must form therefore a redundant set. In other words, given a set of values for $f_{\mu\nu}$, they may not in general be expressible via (1.1.6) in terms of a potential a_{μ}. However, in the particular problem we are considering, the constraint (5.1.11) implies that, for all x not on the monopole world-line $Y(\tau)$, the Bianchi identity is satisfied, which guarantees by the Poincaré lemma, that the gauge potential a_{μ} exists at all these points, thus removing exactly the redundancy of the variables $f_{\mu\nu}$. Hence in this particular problem, it just so happens that we *are* allowed to replace a_{μ} by $f_{\mu\nu}$ as variables, and $f_{\mu\nu}$, being gauge invariant, requires no patching even in the presence of a monopole. This simplifies the problem enormously and makes its solution almost trivial.

Extremizing then $\mathcal{A}'$ with respect to $f_{\mu\nu}(x)$ and $Y^{\mu}(\tau)$, one obtains the Euler–Lagrange equations:

$$f^{\mu\nu}(x) = 4\pi\{\tfrac{1}{2}\epsilon^{\mu\nu\rho\sigma}(\partial_{\sigma}\lambda_{\rho}(x) - \partial_{\rho}\lambda_{\sigma}(x))\}, \qquad (5.1.13)$$

and

$$m\frac{d^2 Y_{\mu}(\tau)}{d\tau^2} = -4\pi\tilde{e}\{\partial_{\nu}\lambda_{\mu}(Y(\tau)) - \partial_{\mu}\lambda_{\nu}(Y(\tau))\}\frac{dY^{\nu}(\tau)}{d\tau}, \qquad (5.1.14)$$

which are to be satisfied together with the constraint equation (5.1.11).

Equation (5.1.13) says just that ${}^*f_{\mu\nu}$ is a gauge field with

$$\bar{a}_{\mu}(x) = 4\pi\lambda_{\mu}(x) \qquad (5.1.15)$$

as the gauge potential, while (5.1.14) can be rewritten using (5.1.13) as:

$$m\frac{d^2 Y^{\mu}(\tau)}{d\tau^2} = -\tilde{e}\,{}^*f^{\mu\nu}(Y(\tau))\frac{dY_{\nu}(\tau)}{d\tau}. \qquad (5.1.16)$$

One sees that the equations (5.1.11) and (5.1.16), which are the equations of motion for a magnetic monopole deduced from its definition as a topological obstruction, are in fact the same as (5.1.4) and (5.1.5) respectively apart from a replacement of e by $\tilde{e}$ and $f^{\mu\nu}$ by ${}^*f^{\mu\nu}$. We conclude therefore that had we started instead with an electric charge e considered as a monopole of the dual field ${}^*f_{\mu\nu}$, we would have arrived at (5.1.4) and (5.1.5) as equations of motion for an electric charge in an electromagnetic field. In other words, we have shown that the intrinsic interaction with the electromagnetic field which an electric charge possesses by virtue of its topological structure is in fact the same as the standard electromagnetic interaction deduced from experiment by Faraday and Maxwell, substantiating thus our earlier claim that the latter interaction can be obtained as a consequence of topology instead of having to be introduced as an additional term (5.1.3) in the action.

What happens in the more interesting case of a quantum particle? The same intuitve arguments as those given for the classical case above will lead also to the conclusion that there will be an intrinsic interaction between the charge and the field due to the dual role of the charge as a monopole. Again, the derivation of this intrinsic interaction is the same independently of whether the charge is electric or magnetic; so for easier reference to earlier material we shall work it out first for the magnetic charge. Supposing that we are dealing with the usual Dirac particle, we would write for the free action of the particle, instead of (5.1.2):

$$\mathcal{A}_M^0 = \int d^4x \bar{\psi}(x)(i\partial_\mu \gamma^\mu - m)\psi(x) \qquad (5.1.17)$$

and for the constraint equation, instead of (5.1.11):

$$\partial_\nu {}^*f^{\mu\nu}(x) = -4\pi \tilde{e}\bar{\psi}(x)\gamma^\mu \psi(x), \qquad (5.1.18)$$

where we have replaced the classical current on the right of (5.1.11) by its standard quantum analogue. The equations of motion are then to be obtained by extremizing the corresponding free action $\mathcal{A}^0$ with respect to the variables ψ and a_μ subject to the constraint (5.1.18), and they will again be coupled equations in ψ and a_μ, representing thus an interacting particle–field system.

At first sight, (5.1.18) may look a little disturbing since the constraint is supposed to be no more than just a covariant representation of the topological definition of the monopole charge, which ought to be quantized. This means that for $\mu = 0$, the integral over any volume of the right-hand side of (5.1.18) should give only values equal to $4\pi\tilde{e}$ times an integer, which it clearly does not for a general $\psi(x)$. For this reason, we believe that for a truly consistent interpretation of (5.1.18), $\psi(x)$ ought to be considered as a second quantized field so that for $\mu = 0$, the right-hand side of (5.1.18) becomes just a numerical constant times the number operator counting the (integral) number of mono-poles occuring at x, whose integral over any volume will then be quantized. In the present context, however, we shall accept (5.1.18) at face value without pursuing further the second quantized interpretation, which would otherwise take us beyond the scope intended.

Let us return now to the action principle with (5.1.17) instead of (5.1.2) and (5.1.18) instead of (5.1.11). The problem of patching with a_μ as variables is still with us, and we would like to adopt the same tactics as that used above in the classical case by changing the variables to $f_{\mu\nu}$. This may seem inadvisable since we know from the Aharonov–Bohm–Chambers experiment that, for describing the electromagnetic interaction of a charged quantum particle, $f_{\mu\nu}$ is not enough — one needs a gauge potential to account for the phase of the complex wave function. However, we recall that the potential one needs for describing a magnetic charge interacting with an electromagnetic field is not the potential a_μ of the Maxwell field $f_{\mu\nu}$ but a potential $\bar{a}_\mu$ of the dual field $^*f_{\mu\nu}$ as defined in (5.1.8). Furthermore, we saw in (5.1.13) and (5.1.14) while solving the variational problem that the Lagrange multiplier λ_μ emerged automatically to play the role of the potential $\bar{a}_\mu$. Although this λ_μ was eliminated there in the end and disappeared from the final equations of motion, it may survive in the equations for the quantum particle and be coupled to the particle wave function as a gauge potential should.

To see whether this will indeed be the case, incorporate the constraint (5.1.18)into an auxiliary action as follows:

$$\mathcal{A}' = \mathcal{A}_F^0 + \mathcal{A}_M^0 + \int d^4x\,\lambda_\mu(x)\{\partial_\nu {}^*f^{\mu\nu}(x) + 4\pi\tilde{e}\bar{\psi}(x)\gamma^\mu\psi(x)\} \qquad (5.1.19)$$

which is to be extremized under unconstrained variations of $f_{\mu\nu}$ and ψ. The Euler–Lagrange equation obtained by extremizing with respect to $f_{\mu\nu}$ remains as (5.1.13), while by extremizing with respect to $\bar{\psi}(x)$ one obtains:

$$(i\partial_\mu\gamma^\mu - m)\psi(x) = -\tilde{e}\bar{a}_\mu\gamma^\mu\psi(x), \qquad (5.1.20)$$

with $\bar{a}_\mu$ defined as in (5.1.15). One notices then that whereas in the classical case λ_μ occurred in the Euler–Lagrange equations only in the combination $(\partial_\nu\lambda_\mu - \partial_\mu\lambda_\nu)$ allowing it to be eliminated from the equations of motion, this is no longer true in the quantum case. Instead, λ_μ remains obligingly behind to serve as the new gauge potential $\bar{a}_\mu$, coupling to the wave function ψ exactly as a potential is expected to do. Together with the constraint (5.1.18), the equations (5.1.13) and (5.1.20) then constitute the equations of motion for a quantum particle carrying magnetic charge $\tilde{e}$ moving in an electromagnetic field.

As in the classical case, one sees that had one gone through a parallel derivation by starting instead with an electric charge considered as a monopole of the dual Maxwell field $^*f_{\mu\nu}$, one would have arrived at the same equations apart from a replacement of the magnetic charge $\tilde{e}$ by the electric charge e, the dual field $^*f_{\mu\nu}$ by $f_{\mu\nu}$ and the potential $\bar{a}_\mu$ by the usual potential a_μ. Explicitly, the equations obtained would be:

$$\partial_\nu f^{\mu\nu}(x) = -4\pi e\bar{\psi}(x)\gamma^\mu\psi(x), \qquad (5.1.21)$$

and

$$(i\partial_\mu\gamma^\mu - m)\psi(x) = -ea_\mu(x)\psi(x), \tag{5.1.22}$$

together with the equation (1.1.6) giving $f_{\mu\nu}(x)$ in terms of the potential $a_\mu(x)$. These are exactly the standard equations of motion of a Dirac particle carrying electric charge e moving in an electromagnetic field, which are conventionally derived by extremizing an action obtained by adding to the free action an interaction term of the form:

$$\mathcal{A}_I = e \int d^4x \bar\psi(x) a_\mu(x)\gamma^\mu\psi(x). \tag{5.1.23}$$

Thus, once again, one sees that one can indeed derive the correct interaction as a consequence of topology without introducing an interaction term in the action. Furthermore, one obtains in this way the interaction *uniquely* without having to make the *minimal coupling hypothesis* which is usually invoked to 'deduce' the conventional interaction term (5.1.23).

An interesting new feature in the quantum theory outlined above which was not apparent in the classical case above is that the equations (5.1.20) and (5.1.13) exhibit a local gauge invariance under a new gauge transformation:

$$\psi \longrightarrow \exp i\tilde{e}\alpha \; \psi, \tag{5.1.24}$$

and

$$\bar{a}_\mu \longrightarrow \bar{a}_\mu + \partial_\mu\alpha. \tag{5.1.25}$$

in addition to the original gauge transformation transforming the gauge potential a_μ. It has emerged as a degeneracy in the solution of the variational problem formulated in terms of $f_{\mu\nu}$ and ψ, both of which are invariant under the original gauge transformation. As a result, the system now possesses a $U(1) \times \widetilde{U(1)}$ local gauge invariance, where $\widetilde{U(1)}$ referring to (5.1.24) and (5.1.25) has in fact the opposite parity to the original $U(1)$ since it is associated with $\bar{a}_\mu$, the potential of $^*f_{\mu\nu}$ which is opposite in parity to $f_{\mu\nu}$. We shall call this phenomenon a *chiral doubling* of the gauge symmetry, to which we shall have occasion to return.

5.2 Duality

In the preceding section we have noted two very special properties of the abelian theory. First, at any point in space-time where the field is source-free, we have $\partial^\nu f_{\mu\nu}(x) = 0$, which may be interpreted as the Bianchi identity for the dual field $^*f_{\mu\nu}$, and hence implies by the Poincaré lemma that the latter field is also a gauge field derivable from a potential as per (1.1.6). Secondly, we know that a monopole of the Maxwell field $f_{\mu\nu}$, defined as a topological obstruction in that field, is equivalent to a source of the dual field ${}^*f_{\mu\nu}$ as shown in (5.1.11) or (5.1.18), and conversely also a monopole of ${}^*f_{\mu\nu}$ is a source of $f_{\mu\nu}$. There is thus a symmetry between the field and the dual field as between also a source and a monopole, which symmetry is commonly known as the *dual symmetry* of electromagnetism.

For a nonabelian Yang–Mills theory, however, neither of these two special properties apply. Although at any source-free point of the field $F_{\mu\nu}$, one still has the Bianchi identity holding for the dual field ${}^*F_{\mu\nu}$, this no longer guarantees by itself, for lack of an equivalent Poincaré lemma, that ${}^*F_{\mu\nu}$ is a gauge field derivable from a potential. The usual Poincaré lemma as stated in Section 3.3 gives conditions on when an ordinary 2-form is the curl of a 1-form, but has nothing to say about *Lie algebra-valued* forms and *covariant* curls. Further, as we have already asserted in Chapter 2, neither can one use Gauss' theorem in the same way as one did in the abelian case to deduce that a monopole charge in the field $F_{\mu\nu}$ is equivalent to a source in the dual field ${}^*F_{\mu\nu}$. For these reasons, one says that there in general is no dual symmetry in nonabelian theories.

Consider next interactions. In the abelian case, because of dual symmetry, we have two different ways to deduce the standard electromagnetic interactions according to whether we choose to regard the charge (either electric or magnetic) as a source or a monopole. In the first choice, we add an interaction term to the free action as in (5.1.3) or (5.1.23); in the second we use a constrained action principle as in (5.1.12) and (5.1.19). The result is the same. In a nonabelian theory, on the other hand, sources and monopoles are in general different objects. Indeed, even their charges take different values; for example, for an $SO(3)$ theory, source charges are group representations such as 'singlets' or 'triplets', while monopole charges are labelled only by a sign. The general procedure adopted in the abelian theory for deriving interactions still applies in principle to each case separately, but they are no longer interchangeable, and should give results which need have no similarity to each other. Let us then examine each in turn.

Consider first the interaction of a source. Instead of (5.1.1) for the free action of the field, we have:

$$\mathcal{A}_F^0 = -\frac{1}{16\pi} \int d^4x \operatorname{Tr}\left(F_{\mu\nu}(x)F^{\mu\nu}(x)\right), \qquad (5.2.1)$$

where the field tensor $F_{\mu\nu}(x)$ is an element of the gauge Lie algebra, in other words a matrix in the internal space indices, and the symbol[1] Tr in (5.2.1)

---

[1] Here the trace is normalized to give $\operatorname{Tr}(F_{\mu\nu}F^{\mu\nu}) = \sum F^i_{\mu\nu}F^{i\mu\nu}$ for i running over the generators of the algebra.

denotes a trace over these indices. For the particle, one can choose to consider again either classical or quantum mechanics. For reasons which will be apparent, we shall first deal with the quantum case, of say a Dirac particle, for which the standard expression for the free action is:

$$\mathcal{A}_M^0 = \int d^4x\, \bar{\psi}(x) \left(i\partial_\mu \gamma^\mu - m\right) \psi(x). \qquad (5.2.2)$$

This is formally the same as (5.1.17) in the abelian case, but now has the wave function $\psi(x)$ in the fundamental representation of the gauge group. Thus, for example, for the gauge group $SU(2)$, $\psi(x)$ will be a 2-spinor in internal space indices. For the interaction term, one usually invokes again the *minimal coupling hypothesis* and replaces the derivative ∂_μ in (5.2.2) by the covariant derivative $\partial_\mu - igA_\mu(x)$, obtaining thereby:

$$\mathcal{A}_I = g \int d^4x \bar{\psi}(x) A_\mu(x) \gamma^\mu \psi(x), \qquad (5.2.3)$$

quite analogous to the abelian case.

The equations of motion are then obtained by extremizing the total action:

$$\mathcal{A} = \mathcal{A}_F^0 + \mathcal{A}_M^0 + \mathcal{A}_I \qquad (5.2.4)$$

with respect to the dynamical variables of the system, in this case the gauge potential A_μ and the particle wave function ψ. The result is of course the well-known Yang–Mills equations:

$$D_\nu F^{\mu\nu}(x) = -4\pi g \bar{\psi}(x)\gamma^\mu \psi(x), \qquad (5.2.5)$$

and

$$(i\partial_\mu \gamma^\mu - m)\psi(x) = -gA_\mu(x)\gamma^\mu \psi(x). \qquad (5.2.6)$$

Surprisingly, the corresponding classical theory is not so easy to formulate as one might think at first sight. The classical equations of motion are known, but only by taking the classical limit of the above quantum equations of motion. The result is known as the Wong equations, namely:

$$D_\nu F^{\mu\nu}(x) = -4\pi g \int I(\tau) \frac{dY^\mu(\tau)}{d\tau} \delta^4(x - Y(\tau)), \qquad (5.2.7)$$

and

$$m\frac{d^2Y^\mu(\tau)}{d\tau^2} = -g\,\mathrm{Tr}\left[I(\tau)F^{\mu\nu}(Y(\tau))\right]\frac{dY_\nu(\tau)}{d\tau}, \qquad (5.2.8)$$

where $I(\tau)$ is an element of the gauge Lie algebra. One notes the similarity in form of these equations to respectively the Maxwell equation (5.1.4) and the

Lorentz equation (5.1.5) for a classical electric charge moving in an electromagnetic field, except for the fact that the charge e which was just a number in the abelian case is replaced by the quantity $gI(\tau)$ which is an element of the gauge Lie algebra. Because of this difference, however, the equations cannot, as far as one knows, be derived from an action principle in the usual way by introducing a scalar interaction term into the action. Indeed, Yang suggested in a private communication to the authors that a generalization of the usual action principle is needed to encompass the classical system described by the Wong equations.

What about monopoles? From the discussion in the last section, we recognize that in electromanetism they are logically as valid a representation of the electric charge as are sources, while the principle for deriving their interactions with the field may even be said to have a superior aesthetic appeal. Thus, in the same spirit of Yang and Mills' original proposal to generalize the dynamics of a source interacting with an abelian field, it ought to be of equal theoretical interest to generalize the dynamics of a monopole to nonabelian theories. The principle laid down for the abelian theory should still apply, but the implementation is a little more complicated, and will be dealt with in the next two sections.

5.3 Classical Theory of Nonabelian Monopoles

From our intuitive arguments above, one expects the equations of motion for a monopole in a nonabelian gauge field to be uniquely determined by the topology as in the abelian theory. However, whereas the equations so obtained in the abelian theory were merely the dual of the familiar Maxwell and Lorentz equations for a source, the equations for a nonabelian monopole will be new and otherwise entirely unknown.

The formulation of the classical problem for a nonabelian monopole is very similar to the abelian case in the last section. One begins by writing down the free action, which is formally the same as (5.1.9), except that $\mathcal{A}_F^0$ is now replaced by the standard Yang–Mills action (5.2.1). $\mathcal{A}_M^0$ remains the same as (5.1.2). As in the abelian theory, equations of motion are to be obtained as Euler–Lagrange equations from extremizing the free action with respect to the gauge potential $A_\mu(x)$ and the particle coordinates $Y^\mu(\tau)$ under the constraint that the particle at $Y^\mu(\tau)$ occurring in $\mathcal{A}_M^0$ should carry a monopole charge of the gauge field $F_{\mu\nu}(x)$.

The solution of the variational problem so posed, however, is technically much more difficult. In the abelian case, our solution relied heavily on two tactics. First, we reduced the topological constraint defining the monopole charge to a differential form (5.1.11) local in space-time. Secondly, to avoid the complexity of patching, we replaced the gauge potential as variables by the field tensor. Neither of these tactics are directly applicable here. Because the electromagnetic concept of flux fail to generalize to the nonabelian theory, as detailed in Section 2.2, we do not know a way for reducing the topological constraint defining the monopole charge to a differential form local in x. Further, the complexity of patching is not avoided by using the field tensor $F_{\mu\nu}$ as variables, since, being only gauge covariant and not invariant as in the abelian case, they are patch-dependent quantities.

In view of the material developed in Chapter 4, a strategy which suggests itself is to adopt loop space variables. First, we have seen that the topological definition of the monopole charge, which was given only abstractly in Chapter 2 as a homotopy class, can be stated explicitly in terms of the loop variables $F_\mu[\xi|s]$, albeit in a rather complicated fashion, as in (4.4.5). Secondly, we recall that loop variables are patch-independent, so that $F_\mu[\xi|s]$ in particular, being closely related to the field tensor according to (4.2.8), promise to be suitable replacements for the gauge potential A_μ as variables, in analogy with $f_{\mu\nu}$ for the abelian case.

A basic difficulty of loop variables is, of course, their high degree of redundancy, which has to be removed. The beauty of the present problem, however, is that the constraints we need to remove the redundancy are implied by those we wish in any case to impose on the variations for deriving the equations of motion, namely the topological conditions (4.4.5) defining the monopole charge. The reader is referred back to the discussion at the end of Section 4.5 for what was called there the extended Poincaré lemma in the presence of monopoles. It is interesting also to compare the present situation with that in the abelian case presented in Section 5.1, where it was again the constraint (5.1.11) defining the monopole charge which removed the redundancy from $f_{\mu\nu}$ as variables. Indeed, these coincidences suggest that loop variables are probably the natural variables to use for nonabelian monopoles.

Adopting then the strategy suggested, let us recast our problem entirely in terms of the variables $F_\mu[\xi|s]$. Using the formula (4.2.8) giving $F_\mu[\xi|s]$ in terms of ordinary field variables, the action $\mathcal{A}_F^0$ in (5.2.1) can be rewritten as:

$$\mathcal{A}_F^0 = -\frac{1}{4\pi\bar{N}} \int \delta\xi \int_0^{2\pi} ds \, \mathrm{Tr}\left\{F_\mu[\xi|s]F^\mu[\xi|s]\right\} \left[\frac{d\xi^\alpha(s)}{ds}\frac{d\xi_\alpha(s)}{ds}\right]^{-1}, \quad (5.3.1)$$

where $\bar{N}$ is a numerical (infinite) normalization factor:

$$\bar{N} = \int_0^{2\pi} ds \int \prod_{s'\neq s} d^4\xi(s'). \quad (5.3.2)$$

Added to $\mathcal{A}_M^0$ in (5.1.2), this gives us the total free action $\mathcal{A}^0$ in terms of the variables $F_\mu[\xi|s]$ and $Y^\mu(\tau)$. To find the equations of motion, $\mathcal{A}^0$ has to be extremized with respect to $F_\mu[\xi|s]$ and $Y_\mu(\tau)$ under the constraint that $Y_\mu(\tau)$ carries a monopole charge.

As we have learned in Chapter 4 above, the defining condition for a monopole charge can be stated in loop space in two equivalent versions, one integral (4.4.5) and the other differential (4.4.6), either of which, can be used as the topological constraint in deriving the equations of motion. Again, the constraints can be incorporated into an auxiliary action by the method of Lagrange multipliers. One has to introduce as many multipliers as there are constraints, so that if one chooses to use the differential version (4.4.6), one has to introduce say $L^{\mu\nu}[\xi|s]$, each $\mu\nu$ component being a functional of the loop ξ and a function of the point s on the loop. One writes then:

$$\mathcal{A}' = \mathcal{A}^0 + \int \delta\xi\, ds\, \mathrm{Tr}\left\{ L^{\mu\nu}[\xi|s]\{G_{\mu\nu}[\xi|s] + 4\pi J_{\mu\nu}[\xi|s]\}\right\}, \tag{5.3.3}$$

where $J_{\mu\nu}[\xi|s]$ denotes just $-1/4\pi$ times the right-hand side of (4.4.6):

$$J_{\mu\nu}[\xi|s] = \tilde{g} \int d\tau\, \kappa[\xi|s]\epsilon_{\mu\nu\rho\sigma} \frac{d\xi^\rho(s)}{ds} \frac{dY^\sigma(\tau)}{d\tau}\, \delta^4(\xi(s) - Y(\tau)). \tag{5.3.4}$$

Note that we need to impose the condition (4.4.6) only for $s = s'$ since for the other values it is automatically satisfied. In (5.3.3) as in what follows, $G_{\mu\nu}[\xi|s]$ denotes $G_{\mu\nu}[\xi; s, s']$ smeared in s' around s. If one chooses to use the integral version (4.4.5) of the constraint, then one needs to introduce as Lagrange multipliers, say Λ_Σ, one for each parametrized surface Σ in spacetime, which will be even more numerous. The auxiliary action incorporating the constraint will then contain an integral over all such surfaces. To obtain the equations of motion in each case, the auxiliary action $\mathcal{A}'$ is to be extremized under unconstrained variations of the field variables $F_\mu[\xi|s]$ and the particle coordinates $Y^\mu(\tau)$.[2] The resulting Euler-Lagrange equations will depend on the Lagrange multipliers $L^{\mu\nu}[\xi|s]$ or Λ_Σ, which will then have to be eliminated to yield the final equations of motion. Although in principle straightforward, the derivation is made complicated by the large number of degrees of freedom and is not very transparent. We shall here only quote the result.

Whether using the integral (4.4.5) or the differential (4.4.6) version of the defining topological constraint, one obtains by extremizing the corresponding

---

[2]To be exact, the variables $F_\mu[\xi|s]$ have still to satisfy the transversality condition (4.2.10) representing reparametrization invariance, which can again be incorporated into $\mathcal{A}'$ by introducing further Lagrange multipliers. However, this constraint is relatively easy and straightforward to handle.

auxiliary action and eliminating the Lagrange multipliers the following equations:

$$\frac{\delta}{\delta\xi^\mu(s)}F_\mu[\xi|s] = 0, \qquad (5.3.5)$$

and,

$$m\frac{d^2Y_\mu(\tau)}{d\tau^2} = -\frac{2\tilde{g}}{\tilde{N}}\int ds\int\delta\xi\,\epsilon_{\mu\nu\rho\sigma}\,\mathrm{Tr}\,[\kappa[\xi|s]F^\nu[\xi|s]]$$

$$\times\frac{d\xi^\rho(s)}{ds}\frac{dY^\sigma(\tau)}{d\tau}\left[\frac{d\xi(s)}{ds}\right]^{-2}\delta^4(\xi(s) - Y(\tau)). \quad (5.3.6)$$

Together with the constraint equation (4.4.6),

$$G_{\mu\nu}[\xi|s] = -\frac{\pi}{g}\int d\tau\,\kappa[\xi|s]\epsilon_{\mu\nu\rho\sigma}\frac{d\xi^\rho(s)}{ds}\frac{dY^\sigma(\tau)}{d\tau}\delta^4(\xi(s) - Y(\tau)),$$

the equations (5.4.8) and (5.3.6) then constitute the complete set of equations of motion for a monopole at $Y(\tau)$ moving in the gauge field $F_\mu[\xi|s]$. We stress once more that, as in the abelian theory, these equations have been derived as a unique consequence of the free action $\mathcal{A}^0$ under the topological condition defining the monopole charge without introducing at any stage an interaction term.

The equations so derived, however, are unfortunately rather unwieldy in that they are expressed in terms of unfamiliar loop variables, the physical significance of which is a little hard to appreciate. It would help in our understanding if we were to recast these equations in terms of ordinary space-time local variables. We can do so using the expressions (4.2.8) and (4.3.15) for respectively the loop quantities $F_\mu[\xi|s]$ and $G_{\mu\nu}[\xi|s]$ derived in the preceding chapter. Strictly speaking, these relations do not apply at the monopole position since, at that particular point local field quantities like $F_{\mu\nu}(x)$ and $A_\mu(x)$, being patched, do not really exist. However, if we ignore for the moment this rather essential complication, we obtain first that the equation (5.3.5) is equivalent to the standard source-free Yang–Mills equation:

$$D_\nu F^{\mu\nu}(x) = 0, \qquad (5.3.7)$$

a fact first noted by Polyakov. That this is so can be seen as follows. Substituting (4.2.8) into (5.4.8) and using the fact deducible from (4.2.9) that

$$\left\{\frac{\delta}{\delta\xi^\mu(s)}\Phi_\xi(s,o)\right\}\Phi_\xi^{-1}(s,o) = igA_\mu(\xi(s)), \qquad (5.3.8)$$

we get:

$$D^\mu F_{\mu\nu}(\xi(s))\frac{d\xi^\nu(s)}{ds} = 0, \tag{5.3.9}$$

which is to hold for all parametrized loops ξ at all points s, and hence for all $d\xi^\nu(s)/ds$. This will be true if and only if the source-free Yang–Mills equation (5.3.7) is satisfied.

Secondly, substituting (4.3.15) into the constraint equation (4.4.6), one obtains:

$$D_\nu{}^*F^{\mu\nu}(x) = -4\pi\tilde{g}\int d\tau\, [\Phi_\xi(s,0)\kappa[\xi|s]\Phi_\xi^{-1}(s,0)]_{\xi(s)=x}\frac{dY^\mu(\tau)}{d\tau}\delta^4(x-Y(\tau)), \tag{5.3.10}$$

where we have again used the fact that the loop space equation (4.4.6) has to be satisfied for all parametrized loops ξ at all points s. Now the quantity inside the square bracket in (5.3.10) can depend on the parametrized loop ξ only through the point $\xi(s)$, and since it is to be evaluated only at $x = Y(\tau)$ by virtue of the δ-function, we shall replace it by the symbol $K(\tau)$. Thus, the equation (5.3.10) becomes:

$$D_\nu{}^*F^{\mu\nu}(x) = -4\pi\tilde{g}\int d\tau\, K(\tau)\frac{dY^\mu(\tau)}{d\tau}\delta^4(x-Y(\tau)). \tag{5.3.11}$$

Finally, introducing the symbol $K(\tau)$ together with (4.2.8) into the equation (5.3.6), we obtain:

$$m\frac{d^2Y_\mu(\tau)}{d\tau} = -\frac{\tilde{g}}{2\bar{N}}\int ds\int \delta\xi\, \epsilon_{\mu\nu\rho\sigma}\mathrm{Tr}[K(\tau)F^{\nu\rho}(\xi(s))]\frac{dY^\sigma(\tau)}{d\tau}\delta^4(\xi(s)-Y(\tau)) \tag{5.3.12}$$

where we have averaged over all directions of the tangent vector $d\xi^\mu(s)/ds$ to the loop ξ at s. Integrating then over ξ and recalling the definition (5.3.2) of the normalization factor $\bar{N}$, we have:

$$m\frac{d^2Y^\mu(\tau)}{d\tau^2} = -\tilde{g}\,\mathrm{Tr}\,[K(\tau)^*F^{\mu\nu}(Y(\tau))]\frac{dY_\nu(\tau)}{d\tau}. \tag{5.3.13}$$

One sees then that of the three equations that we have derived governing the motion of a monopole in a gauge field, two of them, namely (5.3.11) and (5.3.13) are very similar in appearance to the Wong equations for the motion of a classical point source of a Yang–Mills field as quoted in (5.2.7) and (5.2.8). Indeed, if from the Wong equations, one makes the replacements $F_{\mu\nu}(x) \to {}^*F_{\mu\nu}(x)$, $I(\tau) \to K(\tau)$, and $g \to \tilde{g}$, one obtains exactly the equations (5.3.11) and (5.3.13). Further, the remaining equation we have derived, namely (5.3.7), is no more than just the Bianchi identity for the dual Yang–Mills field $^*F_{\mu\nu}$. In other words, the equations of motion we have derived for the nonabelian monopole look rather like the dual transforms of the Wong equations for the motion of a source. There is however one very important difference, namely that, very intriguingly, the covariant derivative D_μ in equation (5.3.11) is still defined via (2.2.9) in terms of the potential A_μ, and *not* in terms of some potential, $\bar{A}_\mu$ say, of the dual field ${}^*F_{\mu\nu}$ as one might naively expect. Indeed, as explained above in the preceding section, such a potential $\bar{A}_\mu$ does not exist in general.

Although the dual symmetry between the above equations for the non-abelian monopole and the Wong equations for a source are not exact, the quasi-dual relationship between them suggests that the dynamics have some similarities in the two cases. That this can be so is already a great surprise, for in the nonabelian theory, monopoles and sources started out as very different objects. We recall that even their charges are very different, with monopole charges labelled by homotopy classes but source charges by representations of the gauge group. In formulating the dynamics also, we have followed entirely different routes for the two types of charges. Yet, for some intriguing reason, the result seems still to bear a close similarity to each other.

5.4   Dual Yang–Mills Theory

As an example of the quantum theory of nonabelian monopoles, let us consider in parallel with the Yang–Mills theory of a point source a point Dirac particle carrying a monopole charge moving in a nonabelian gauge field. The free actions of the field and particle remain then the same as (5.2.1) and (5.2.2). Instead of introducing an interaction term into the action such as (5.2.3) for a source, interactions between the field and the monopole will come about automatically by virtue of the topological constraint defining the monopole charge. Our first task is therefore to write down this constraint explicitly in terms of our chosen dynamical variables.

For reasons already made clear above while solving the analogous problem of the classical nonabelian monopole, we shall work in loop space and seek to write the defining topological constraint for the monopole charge in a loop differential form similar to (4.4.6). As explained above also, the condition (4.4.6) is formally equivalent to the condition (5.3.10) local in space-time where the right-hand side represents a sort of nonabelian monopole current similar to the magnetic current of (5.1.11) for the abelian case. In the abelian quantum problem, we succeeded in solving our problem by substituting for the classical the quantum current as per (5.1.18). This suggests that we replace also the right-hand side of (5.3.10) by a quantum current, say perhaps:

$$D_\nu {}^*F^{\mu\nu}(x) \stackrel{?}{=} -4\pi\tilde{g}\,(\overline{\tilde{\psi}(x)}\,\gamma^\mu\,T_i\,\tilde{\psi}(x))\,\tau^i, \qquad (5.4.1)$$

where $\tilde{\psi}(x)$ represents the wave function of the monopole and τ^i a generator of the gauge Lie algebra. The quantum 'monopole current' on the right of (5.4.1) would transform like an element of the algebra as it should if the monopole wave function $\tilde{\psi}(x)$ transforms like a representation of the algebra and T_i are matrices representing the generators in that representation. This suggestion, however, is not quite correct. Just as in the abelian case treated in Section 5.1 where a monopole though magnetically charge is electrically neutral and therefore transforms not under the original $U(1)$ but only under a new $\tilde{U}(1)$ transformation, so now a monopole in a nonabelian theory should also be *colour* neutral and transform as a representation not under the original Yang–Mills U but under some new $\tilde{U}$-transformations. Hence, as it stands, the equation (5.4.1) is inconsistent, having the left-hand side covariant but the right-hand side invariant under U-transformations, while under $\tilde{U}$-transformations, the left-hand side is invariant but the right-hand side covariant.

We can correct the above discrepancy in (5.4.1) as follows. Introduce at each space-time point x two local frames for the gauge Lie algebra, one transforming under U- and the other under $\tilde{U}$-transformations. Let $\omega(x)$ be the matrix which transforms from the U-frame to the $\tilde{U}$-frame at x. Then simultaneously under a U-transformation:

$$\psi(x) \longrightarrow (1 + ig\Lambda(x))\psi(x), \qquad (5.4.2)$$

and a $\tilde{U}$-transformation:

$$\tilde{\psi}(x) \longrightarrow (1 + i\tilde{g}\tilde{\Lambda}(x))\tilde{\psi}(x), \qquad (5.4.3)$$

the matrix $\omega(x)$ will transform as:

$$\omega(x) \longrightarrow (1 + i\tilde{g}\tilde{\Lambda}(x))\omega(x)(1 - ig\Lambda(x)). \qquad (5.4.4)$$

Instead of (5.4.1), we then write:

$$D_\nu {}^*F^{\mu\nu}(x) = -4\pi\tilde{g}\,[\overline{\tilde{\psi}(x)}\omega(x)\gamma^\mu T_i\omega^{-1}(x)\tilde{\psi}(x)]\,\tau^i, \qquad (5.4.5)$$

which will now have correct transformation properties under both U- and $\tilde{U}$-transformations, being covariant under the first and invariant under the second type of transformations.

Returning now to loop space, we then write in analogy with (4.4.6) the defining constraint for the monopole charge as:

$$G_{\mu\nu}[\xi|s] = -4\pi\tilde{g}\int d^4x\,\epsilon_{\mu\nu\rho\sigma}\overline{[\tilde{\psi}(x)}\Omega_\xi(s,0)\gamma^\rho T_i\Omega_\xi^{-1}(s,0)\tilde{\psi}(x)]\tau^i\frac{d\xi^\sigma(s)}{ds}\delta^4(x-\xi(s))$$
$$(5.4.6)$$

where

$$\Omega_\xi(s,0) = \omega(\xi(s))\Phi_\xi(s,0) \qquad (5.4.7)$$

is the operator for parallel phase transport from the U-frame at the reference point $\xi_0 = \xi(0) = \xi(2\pi)$ to the $\tilde{U}$-frame at $\xi(s)$ along the loop parametrized by ξ. The constraint (5.4.6) is then what we wish to impose on variations in the dynamical variables when extremizing the (free) action to deduce the (interacting) equations of motion of the field–monopole system.

Incorporating the constraint using the method of Lagrange multipliers, we again construct the auxiliary action as in (5.3.3) with only the difference that $J_{\mu\nu}[\xi|s]$ now refers to the quantum analogue, namely $-1/4\pi$ times the right-hand side of (5.4.6). As in the classical problem, the field action $\mathcal{A}_F^0$ is expressed in terms of the loop variables $F_\mu[\xi|s]$ as (5.3.1) and equations of motion are to be obtained by extremizing $\mathcal{A}'$ with respect to $F_\mu[\xi|s]$ and $\tilde{\psi}(x)$. Again, we shall just quote the result:

$$F^\mu[\xi|s] = -4\pi\bar{N}\left(\frac{d\xi(s)}{ds}\right)^2\left\{\frac{\delta}{\delta\xi^\nu(s)}L^{\mu\nu}[\xi|s] - ig[F_\nu[\xi|s], L^{\mu\nu}[\xi|s]]\right\}, \quad (5.4.8)$$

which implies (5.3.5), and

$$(i\partial_\mu\gamma^\mu - m)\tilde{\psi}(x) = -\tilde{g}\tilde{A}_\mu(x)\gamma^\mu\tilde{\psi}(x), \qquad (5.4.9)$$

with:

$$\tilde{A}_\mu(x) = 4\pi\int\delta\xi\,ds\,\epsilon_{\mu\nu\rho\sigma}\Omega_\xi(s,0)L^{\rho\sigma}[\xi|s]\Omega_\xi^{-1}(s,0)\frac{d\xi^\nu(s)}{ds}\delta^4(x-\xi(s)). \quad (5.4.10)$$

Together with the constraint equation (5.4.6), (5.4.8) and (5.4.9) constitute the complete set of equations governing the motion of the monopole in the nonabelian Yang–Mills field.

Notice that here, as in the abelian theory for a quantum particle, the Lagrange multipliers cannot be eliminated, but remain behind to act as 'dual potential' $\tilde{A}_\mu$, making the equation (5.4.9) appear as exactly the dual of the Yang-Mills equation (5.2.6) for a source. However, in contrast to the abelian theory, this $\tilde{A}_\mu$ is not a genuine potential in the usual sense and bears a relationship to the dual field tensor ${}^*F_{\mu\nu}$ different from that between A_μ and $F_{\mu\nu}$. Further, analogously to the abelian theory, it can be shown that these equations are invariant under the transformation (5.4.3), accompanied by a corresponding transformation in $L^{\mu\nu}[\xi|s]$ which we shall not exhibit, giving thus again a *chiral doubling* of the original symmetry.

The physical properties of the above monopole equations have not been thoroughly investigated but are in any case not particularly relevant for our present discussion on concepts. It suffices for us just to remind the reader that these equations are derived, as in all the earlier cases considered, from the defining condition of the monopole charge without the introduction into the action of an interaction term.

5.5 Remarks

In the preceding sections, we have shown how the topological definition of a monopole charge suffices by itself to determine the monopole–field interaction and leads to a unique set of equations of motion in each of the classical and quantum mechanical cases examined. Although we have given the derivation in detail only for a single monopole, it ought to be obvious that the arguments are immediately generalizable to a system with any number of monopoles carrying each any charge and moving together in the gauge field. Indeed, it seems possible to extend the formulation even further into a second quantized theory in which the number of monopole charges becomes indefinite. Such a formulation, however, is yet far from complete, and would in any case take us beyond the scope of this present volume restricted to classical fields. Moreover, although the treatment above was explicit only for a theory with gauge group $SO(3)$, it is also readily generalizable to theories with other gauge groups; the details may differ, but not the principle[3]. Since, as we may recall, even the existence of monopole charges as topological obstructions is an intrinsic property of gauge theories, this means that once given some gauge theory, a theory of interacting fields and charges can be developed without further dynamical input. That being the case, would one not perhaps, in the memorable words of Dirac, 'be surprised if Nature should make no use of it'?

Do monopoles then exist? Knowing now the explicit interactions of monopoles, as one claims to have done from the considerations above, one should be able in principle to devise means for detecting them by experiment. There is even an intriguing possibility that some of the particles we already know are in fact monopoles rather than sources as commonly believed. For electromag-

---

[3]In cases where the gauge group is semi-simple, as in the standard electroweak theory, a particle can carry simultaneous monopole charges appropriate to each of the simple 'factor groups' and interact with the individual gauge fields accordingly. For an explicit example of this, see Chan, Scharbach and Tsou (1986).

netism, of course, this last proposition is not particularly meaningful, for in the abelian theory the dynamics of monopoles and sources are in fact the same because of dual symmetry, merely represented in two equivalent descriptions. So long as we know only the electric but not the magnetic charge, then we are at liberty to consider that charge either as a source or as a monopole. For nonabelian theories, however, sources and monopoles are different objects with different dynamics, and yet both are generalizations of the electric charge. It is then a meaningful and legitimate question to ask whether the particles we see in nature are in fact nonabelian sources or monopoles. The conventional wisdom, of course, is that they are all sources, obeying the dynamics prescribed by the Yang–Mills theory through the equations (5.2.5) and (5.2.6), and this is already substantiated by quite a large body of experimental evidence. Nevertheless, without first studying monopole dynamics in more detail, it cannot be excluded that some, or perhaps even all, of the particles in Nature are in fact monopoles which, as indicated above, do seem to have some aesthetic advantages.

At first sight, it may look impossible for sources and monopoles in nonabelian theories to be confused, for even the values of their charges are different. In QCD, for example, sources have charges labelled by $SU(3)$ representations, e.g. triplets, whereas the charges of monopoles are labelled only by a phase, $\exp(2in\pi/3)$. But, as we have seen on closer examination, their dynamics do not appear to be that very different.

However, since our interest here is only in concepts, let us be content with just posing again for the nonabelian theory the question first asked by Dirac in electromagnetism, namely whether monopoles actually occur in Nature, but not making any attempt to answer it.

Chapter 6

Generalized Gauge Structures

6.1 Preamble

The selection of topics considered in the preceding chapters, though far from exhaustive, is we hope enough to illustrate the amazing richness in physical concepts contained in the simple stipulation of local gauge invariance. That being the case, it is doubtless of theoretical interest to speculate on possible generalizations of gauge concepts to a wider context, and it is our purpose in this chapter to consider some of these generalizations in a particular direction. One may object to such a course of action in principle on the grounds that, as far as is known, no physical phenomenon explicitly demands these generalizations, and therefore as physicists, we ought not to indulge in speculations on hypothetical situations. On the other hand, one may counter by the argument that, our vision being restricted by limited knowledge and understanding, it may be that there are already physical circumstances requiring a formulation in a generalized gauge framework without us recognizing the fact, and that we shall be able to do so only when we have these generalizations in hand and have fully understood their consequences. The following historical fact might be quoted as an example in support of this last argument.

Electromagnetism is the first gauge theory known to physicists, and Yang–Mills theory is a generalization in which the abelian gauge group $U(1)$ of electromagnetism is replaced by an arbitrary, not necessarily abelian group. When first proposed in the fifties, it was criticized by many as being of little practical value, since the theory contained several vector boson fields with zero mass, whereas experimentally, only one massless vector boson, namely the photon, was known to exist. Since that time, however, we have grown a little wiser. We now realize that under certain circumstances, what were massless vector bosons can acquire a nonvanishing mass via the Higgs–Kibble mechanism, while in some other circumstances, the vector bosons, though remaining massless, can be permanently confined so as to escape direct experimental detection. As a result, Yang–Mills theory, which was once regarded, at least by some, as a mere mathematical curiosity, has become now the *standard theory* on which all present day physics is based.

Perhaps because of this experience, physicists have grown much more receptive to generalizations of gauge concepts with no immediately experimental applications. Indeed, we may have gone too far the other way, being often engaged in generalizations merely for generalization's sake, with no clear purpose of applying them to physics even in the future, however remote. For this reason, though recognizing the likely importance of generalized gauge concepts to future developments of physics, we shall include in our selection which follows only those generalizations for which at least some circumstances are known where they figure in actual physical theories or in attempts at such. We shall try also to give explicit examples in which these generalizations occur.

From the considerations in Chapter I, one can already discern two basic ingredients in the standard formulation of electromagnetism, namely, first, 4-dimensional Minkowski space-time $\mathbb{R}^4$ and second, the gauge group $U(1)$. It is from the interplay between these two ingredients that one obtains the concept of local gauge symmetry. In mathematical language, as explained in Chapter III, one is dealing with a fibre bundle with $\mathbb{R}^4$ as the base-space, and $U(1)$ as the structure group. Yang–Mills theory generalizes electromagnetism in its second ingredient, namely the symmetry as embodied in the gauge or structure group, by replacing $U(1)$ by an arbitrary group, while leaving the first ingredient, namely space-time or the base space of the bundle, unchanged. In this Chapter, we shall begin by considering some generalizations in concept to the first ingredient, i.e. the base-space of the bundle, which means in physical terms, generalizations to the concept of locality.

One familiar generalization of this sort is the replacement of flat Minkowski space-time by a curved manifold, as is done in the Einstein–Maxwell theory. As far as electromagnetism is concerned, there is in this generalization no really profound change in concept. What is profound in extending from the Maxwell to the Einstein–Maxwell theory is the change in our conception of space-time itself, not in its relationship to the Maxwell field. True, the theory of relativity can itself be considered as a local gauge theory, but as such it requires a specialization rather than a generalization of the gauge concepts. In electromagnetism, as in the Yang–Mills theory, the internal symmetry group, or the fibre of the bundle, need have no relationship with the base manifold, whereas in Einstein's relativity, when considered as a gauge theory, the fibre of the bundle is the tangent space to the base manifold itself. It is this additional structure which gives Einstein's theory its special features not found among gauge theories in general. As the study of these special features lies outside the scope of the present volume, we shall not consider the Einstein–Maxwell theory any further.

The generalizations we shall consider in this chapter, on the other hand, will lead us towards a changed conception of locality eventually to such an extent that it need no longer be associated with points in a space, resulting in what was nicknamed by von Neumann as *pointless geometry*. It will also end up in such a way that the two ingredients of gauge theories which we distinguished, namely space-time and internal symmetry, become no longer disjoint but intimately mixed. The changes in physical concepts involved are often subtle, and in order to elucidate them as best we can, we propose to proceed in several stages starting with generalizations closer to the standard theory with which we are familiar. We shall also illustrate the changes at every stage with concrete examples.

6.2 Group Manifold as Base Space

Gauge theories over flat space-time is in a sense a degenerate case in that the space $\mathbb{R}^4$ is the same as its own translation group. By this we mean the following: any point in space-time $\mathbb{R}^4$ can be obtained by applying a 4-dimensional translation to the origin, and conversely, any 4-dimensional translation applied to the origin specifies a point in space-time $\mathbb{R}^4$. Thus, instead of thinking of Yang–Mills theory as a gauge theory over space-time $\mathbb{R}^4$, one could equally consider it as a gauge theory over the translation group manifold, and hence regard the gauge potential $A_\mu(x)$, not as a parallel transport of phases from point to point in space-time, but rather as a prescription of how phases change under translations through parallel transport.

The translation group is, of course, abelian. Suppose now, in a spirit similar to the original suggestion of Yang and Mills to replace the abelian gauge group of electromagnetism by a nonabelian group, we replace the abelian translation group of $\mathbb{R}^4$ by a nonabelian group, and consider a gauge theory over its manifold. The base space is then no longer flat, i.e. it must have either curvature or torsion, but the idea of a gauge theory over a nonflat base space is already familiar in general relativity, as we have explained in the last section, and presents no difficulty in its implementation. Indeed, if we introduce coordinates (which may require patching), say ξ^μ, to label the points of the group manifold, then we may define as before a gauge potential $A_\mu(\xi)$ to specify parallel transport of phases from any point ξ of the manifold to a neighbouring point $\xi + d\xi^\mu$. The transformation law (1.2.7) or (1.2.8) of A_μ under local gauge transformations, as well as the expression (1.2.12) for the field tensor $F_{\mu\nu}$ in terms of A_μ, would remain valid with simply x replaced by ξ. However, such a formulation of the theory depends on the coordinate system and obscures the fact that the base space is itself a group. For our present purpose, it is more instructive to formulate the theory in another way so that the group structure of the base space becomes manifest. Although this represents merely a recast of what is already known, we shall examine the consequences in some detail since many of them are applicable also to later generalizations.

The fact that the base space is itself a group means that each point in it corresponds to a group element, so that a displacement there from one point to another may be represented by the action, say by left multiplication, of some element of the group. In particular, the displacement from any point in that space to a neighbouring point is representable as the action of a group element differing infinitesimally from the identity, or in other words as the action of an element of the Lie algebra. Similarly, the gauge potential which gives parallel phase transport from point to neighbouring point can alternatively be thought of as a prescription for parallelly transporting phases under the action of a (noncommutative) *displacement algebra*, which generalizes the concept of translations in ordinary Yang–Mills theory over $\mathbb{R}^4$.

Let d_l denote the generators of this displacement algebra satisfying the commutation relations:

$$[d_l, d_k] = C_{lk}^m d_m. \qquad (6.2.1)$$

Multiplication from the left by $(1 + \epsilon^l d_l)$ for infinitesimal ϵ^l displaces any point ξ on the manifold to a neighbouring point ξ', and changes a wave function or field ψ over the manifold to:

$$\psi(\xi') = \psi(\xi) + \epsilon^l d_l \psi(\xi). \qquad (6.2.2)$$

Further, to specify parallel transport of phases under the displacements d_l, one can introduce as before the gauge potential A_l; one says that the phase of ψ at ξ is parallel to its phase at $\xi' = (1 + \epsilon^l d_l)\xi$ if their local values at the two points differ by an amount $g\epsilon^l A_l(\xi)$.

So far, the only difference with the ordinary Yang-Mills theory in flat space is just the formal substitution of the partial derivative ∂_μ by the dispacement operator d_l. Thus, by following closely the arguments in Section 1.2, one easily deduces that under an infinitesimal gauge transformation parametrized by $\Lambda(\xi)$, one obtains for the gauge potential A_l the transformation law, generalizing (1.2.8):

$$A_l' = A_l + [d_l, \Lambda] + ig[\Lambda, A_l]. \qquad (6.2.3)$$

One need only remember that the gauge transformation at the displaced point

$\xi' = (1 + \epsilon^l d_l)\xi$ is given by the operator:

$$(1 + \epsilon^l d_l)(1 + ig\Lambda(\xi))(1 - \epsilon^l d_l). \tag{6.2.4}$$

Equivalently, in terms of the *covariant derivative* defined as:

$$D_l = d_l - igA_l, \tag{6.2.5}$$

we have:

$$\delta A_l = [D_l, \Lambda]. \tag{6.2.6}$$

The formula for the field tensor, however, will be rather different. In the familiar case of a flat base space, the component $F_{\mu\nu}$ of the field tensor (1.2.12) at the point x was $-1/g$ times the phase change obtained by parallel transport around the infinitesimal parallelogram of Figure 1.2 on the $\mu\nu$-plane. The same assertion, however, no longer holds here because parallelograms here do not in general close or, in mathematical terms, because the base manifold has nonzero torsion. By this we mean that by making first a displacement d_l followed by a displacement d_k, we do not always arrive at the same point on the manifold as by making first a displacement d_k followed by a displacement d_l, since by (6.2.1), d_l and d_k do not commute. Indeed, by displacing first by an amount ϵ in direction l and then by an amount ϵ' in direction k, one obtains for the displaced wave function:

$$\{(1 + \epsilon d_l)(1 + \epsilon' d_k)(1 - \epsilon d_l)\}(1 + \epsilon d_l)\,\psi, \tag{6.2.7}$$

while the same displacements in opposite order yield:

$$\{(1 + \epsilon' d_k)(1 + \epsilon d_l)(1 - \epsilon' d_k)\}(1 + \epsilon' d_k)\,\psi. \tag{6.2.8}$$

The difference is:

$$\epsilon\epsilon'\{d_l d_k - d_k d_l\}\psi, \tag{6.2.9}$$

which according to (6.2.1) is $\epsilon\epsilon' C^m_{lk} d_m$ and in general nonzero. This extra piece has to be subtracted from the parallelogram to make it into a closed circuit. Hence, in evaluating the field tensor along the lines described in Section 1.2, we shall need to add to the phase change due to parallel transport around the parallelogram $ABCD$ in Figure 1.2 the change due to parallel transport under the displacement $-\epsilon\epsilon' C^m_{lk} d_m$ which is necessary to close the parallelogram, namely the amount $-g\epsilon\epsilon' C^m_{lk} A_m$, giving:

$$F_{lk} = [d_k, A_l] - [d_l, A_k] + ig[A_l, A_k] + C^m_{lk} A_m, \tag{6.2.10}$$

which is the new formula for the curvature or field tensor in place of (1.2.12). Being by construction a difference between two phases at the same point,

(6.2.10) is guaranteed to be gauge covariant, as was the familiar field tensor (1.2.12) over flat base space. The same conclusion, of course, can also be reached via (6.2.6) by direct computation.

In terms of the covariant derivative D_l in (6.2.5), we may rewrite (6.2.10) as:

$$ig F_{lk} = [D_l, D_k] - C^m_{lk} D_m. \tag{6.2.11}$$

Equivalently, in the notation of differential forms introduced in Section 3.3, we may write:

$$F = dA + ig AA, \tag{6.2.12}$$

where A is the potential 1-form:

$$A = A_l \eta^l, \tag{6.2.13}$$

F is the curvature 2-form:

$$F = \tfrac{1}{2} F_{lk} \eta^l \eta^k, \tag{6.2.14}$$

d is the exterior derivative:

$$d = [d_l, \ ] \eta^l + \tfrac{1}{2} C^m_{lk} \eta^l \eta^k \frac{\partial}{\partial \eta^m} \tag{6.2.15}$$

with an appropriate additional torsion term, and η^l are anticommuting differentials analogous to dx^μ (or left invariant 1-forms in the notation of Section 3.3). Both the expressions (6.2.11) and (6.2.12) will be useful for later reference.

Apart from some additional terms due to torsion in the base, the above development of the theory is entirely analogous to ordinary Yang–Mills theory over flat space-time. Indeed, with the covariant derivative (6.2.5) acting, for example on ψ, and the gauge covariant field tensor F_{lk}, one can proceed as in ordinary Yang–Mills theory to construct a gauge invariant action and formulate the corresponding dynamics. Lacking a particular physical motivation, however, we shall refrain from continuing further in this direction.

One notes that the formulation above made no reference to a coordinate system in the base manifold. To make contact with the coordinate dependent formulation mentioned at the beginning of the section, let us expand the coordinates of the displaced point $\xi' = (1 + \epsilon^l d_l)\xi$ about the original point ξ, thus

$$\xi'^\mu = \xi^\mu + \epsilon^l e^\mu_l, \tag{6.2.16}$$

where the coefficients e^μ_l of the expansion, which depend in general on ξ, may be thought of as a set of local frame vectors giving the directions l of the displacements due to d_l in terms of the directions of the local coordinate axes

μ. Clearly then, the action of the displacement d_l acting on a wave function or field $\psi(\xi)$ is simply:

$$\epsilon^l d_l \psi(\xi) = \psi(\xi') - \psi(\xi) = \epsilon^l e_l^\mu \frac{\partial}{\partial \xi^\mu} \psi(\xi). \tag{6.2.17}$$

In other words, $d_l = e_l^\mu \partial_\mu$ is just the derivative in the direction l. However, the local frame vectors e_l^μ themselves being functions of the coordinates ξ, the operators d_l do not in general commute, but satisfy rather the relations (6.2.1), as can be checked by direct computations. Further, one can similarly verify that F_{lk} in (6.2.10) is just $e_l^\mu e_k^\nu F_{\mu\nu}$, with $A_l = e_l^\mu A_\mu$.

The alternative, coordinate independent, treatment presented above did not, of course, lead to a theory other than the standard gauge theory over a nonflat manifold, but just recast the theory in a different form. The purpose was to emphasize the group structure of the base and to highlight the fact that the concept of local gauge invariance can be understood in two ways. It can be thought of either as invariance under independent phase rotations at different points in the base manifold, or as invariance under independent phase rotations before and after the action of elements of the displacement algebra. Similarly, the gauge potential A can be regarded either as parallel transport of phases from point to neighbouring point on the base manifold, or else as parallel transport of phases under the action of the displacement algebra. This dual interpretation exists already in ordinary Yang–Mills theory over a flat space-time but tends to be overlooked there since the same formulae apply in either case. Here, where the displacement algebra is noncommutative, the two interpretations lead to formulae differing in appearance but equivalent in content. Later on, however, we shall be dealing with situations where the distinction between the two interpretations makes a material difference.

As a physical example of how a gauge structure over a nonabelian group manifold may arise, consider a molecule or atom with a deformed nucleus. We shall regard the nucleus as heavy so that its motion can be treated adiabatically. Let the orientation of the nucleus be specified by Euler angles $\xi^\mu, \mu = 1, 2, 3$, or equivalently by a spatial rotation, ξ say, relative to a fixed coordinate system, or in other words an element of the 3-dimensional rotation group. The whole system of nucleus plus its cloud of electrons can then be described by a Hamiltonian $H(\xi)$ which is an operator in the electronic degrees of freedom but depends on the group element ξ as a parameter.

For each choice of the element ξ, we may seek solutions to the equation:

$$H(\xi) |n; \xi\rangle = E_n(\xi) |n; \xi\rangle, \tag{6.2.18}$$

where $|n; \xi\rangle$ represents a normalized eigenvector of $H(\xi)$. Let us assume for simplicity that the level n is nondegenerate. Then $|n; \xi\rangle$ is specified up to a phase, which, not being an observable quantity, can be chosen arbitrarily and independently for each ξ. Hence, to each point ξ in the rotation group, there is associated a circle representing the values between 0 and 2π that this phase can take, leading thus to a gauge structure of the type considered above in this section, or in mathematical terms, a $U(1)$-bundle over the rotation group manifold. The choice of a vector $|n;\xi\rangle$ with a definite phase for each ξ is in this language then a gauge choice which gives for every ξ a reference phase to which other phases can be referred.

Suppose now one slowly rotates the nucleus by a rotation $\xi(t)$ depending on time t. As t varies, $\xi(t)$ traces out a path in the rotation group. Take now any wave function ψ describing our system. According to standard quantum mechanics, ψ will evolve with time according to the time-dependent Schrödinger equation:

$$i\frac{\partial\psi}{\partial t} = H(\xi(t))\,\psi. \qquad (6.2.19)$$

In particular, starting with an eigenstate ψ_n of $H(\xi(0))$ with the nondegenerate eigenvalue $E_n(\xi(0))$ at time $t = 0$, the adiabatic theorem says that so long as the rotation is slow enough to guarantee that the variations in H are much smaller than its level spacing, (6.2.19) implies that ψ_n will remain an eigenstate of $H(\xi(t))$ at a later time t. Further, the equation specifies also the phase of $\psi_n(t)$ at any later time t. In other words, it can be thought of as defining a transport of phases all along the path in rotation group space specified by $\xi(t)$.

Relative to the reference phase choice $|n;\xi\rangle$ of (6.2.18) above, $\psi_n(t)$ can be written as:

$$\psi_n(t) = e^{-i\int_0^t dt' E_n(t')} e^{i\gamma_n(t)} \,|n;\xi(t)\rangle, \qquad (6.2.20)$$

where the phase in the first exponential factor in (6.2.20) is known as the *dynamical phase*, and:

$$\gamma_n(t) = \int_0^t dt' \langle n;\xi(t')|\frac{d}{dt'}\,|n;\xi(t')\rangle, \qquad (6.2.21)$$

is the adiabatic or *Berry phase*.

In terms of the Euler angles ξ^μ (coordinates of the point ξ on the rotation group manifold), (6.2.21) above can be rewritten as:

$$\gamma_n(t) = \int_0^t d\xi^\mu \langle n;\xi|\frac{\partial}{\partial\xi^\mu}|n;\xi\rangle. \qquad (6.2.22)$$

The quantity $\langle n;\xi(t)|\partial/\partial\xi^\mu|n;\xi\rangle$ specifies how phases are carried or *parallelly transported* from point to neighbouring point in the rotation group manifold, and can thus be interpreted as a gauge potential or a connection for our $U(1)$-bundle over the rotation group manifold. It is easily seen that under a change in the choice of the phase of $|n;\xi\rangle$ by an amount $\alpha(\xi)$, which by (6.2.18) is permissible, $\langle n;\xi|\partial/\partial\xi^\mu|n;\xi\rangle$ will change by $\partial\alpha/\partial\xi^\mu$ as a potential or connection should. It is thus not a physically measurable quantity, since the choice of the phase of $|n;\xi\rangle$ is arbitrary; nor is its integral (6.2.22) over any open path. However, when the integral (6.2.22) is taken over a closed path in the rotation group, it becomes the difference between two phases at the same point in group space, or in physical terms, at the same orientation of the deformed nucleus, in which case it becomes a measurable quantity. It is this that makes the Berry phase so interesting. In particular, the Berry phase taken around an infinitesimal parallelogram in ξ-space can be interpreted as the curvature.

Alternatively, instead of considering the parallel transport of phases from point to neighbouring point on the rotation group manifold using the Euler angles as coordinates, we can consider instead parallel transport under the rotation algebra with generators $d_l = Y_l, l = 1, 2, 3$ for Y_l as introduced in Section 3.2, which satisfy the commutation relations (6.2.1) for $C_{lk}^m = \epsilon_{lkm}$. One can then use the formalism above to construct the curvature etc.

Such a 'cranking model' of systems containing deformed nuclei with its attendant gauge structure has in fact been considered in the literature, though using usually a different language. One notes, however, that in these situations, one usually specifies the Hamiltonian of the system, which specifies in turn via the Schödinger equation the 'gauge potential' $\langle n;\xi|\partial/\partial\xi^\mu|n;\xi\rangle$. On the other hand, in usual gauge theories, the potential A_μ is a dynamical variable whose value is to be determined by the dynamics, e.g. by extremizing the action or solving an equation of motion. Therefore, to complete the analogy, one would need to imagine a situation where the Hamiltonian is not entirely known but has to be determined by extremizing some functional of the 'curvature' constructed from the Berry phase over infinitesimal closed circuits. We have not, however, come across physical situations where it is profitable to do so.

6.3 Distinguishing Two Concepts of Locality

As pointed out at the beginning of the preceding section, ordinary Yang–Mills theory in flat space-time is a degenerate special case in that the base space is identical to its own translation group. That this coincidence is inessential for the formulation of a local gauge theory is obvious since already in the familiar Einstein–Maxwell theory of electromagnetism over a curved Riemannian manifold, there is no such coincidence. In our consideration above of gauge theories over nonabelian group manifolds, we have kept this degeneracy in concepts. However, now that we have familiarized ourselves with the idea of noncommuting displacements, we shall discard this degenerate luxury and proceed to consider gauge structures with their base space different from their group of displacements.

If one examines in detail the arguments of the last section — which have, in fact, deliberately been constructed in this way — one will find little that depends on the identification of the base space with the group generated by the displacement algebra. The only statement which relies on that condition was the assertion that the displacement from any one point to any other point in the base space can be obtained by the repeated action (in that case, left translation) of the algebra elements. However, even on retracting this condition, we can still keep the notion that A_l provides parallel phase transport for whatever displacements which can be effected by elements of the algebra. Furthermore, local gauge invariance can still mean invariance under arbitrary independent phase rotations at different points in the base space. The only difference is that the potential A_l can now provide parallel transport only along paths or orbits traced out by repeated action of the displacement algebra, which need no longer comprise all paths in the base space. Otherwise, all formulae from (6.2.1) to (6.2.17) remain valid, although of course the generalized theory can no longer be regarded as just an ordinary gauge theory over a non-flat manifold as that in the section above.

An example of such a generalized gauge structure arising in physics is provided by the study of anomalies in the quantum theory of chiral fermions moving in an external Yang–Mills field. As elucidated, for example, by Fadeev among others, a wave function in the Schrödinger picture here of the whole system comprising both the quantized Yang–Mills and chiral fermion fields can be regarded as a functional of the Yang–Mills field on $\mathbb{R}^3$ taking values in the Hilbert space, Γ say, of the fermion. We may thus regard the wave function also as a field taking values in Γ over the space of all Yang–Mills potentials in $\mathbb{R}^3$, a space we shall henceforth denote as $\mathcal{A}^3$. This field over $\mathcal{A}^3$ will have internal degrees of freedom as represented by its orientations in Γ, the particular choice of which we shall refer to in future as its 'phase'. Hence we can set up a new bundle structure, beyond the original Yang–Mills bundle structure, with again Γ as fibre but now $\mathcal{A}^3$ as base space. Further, we may regard the components $A_i^a(x)$ of the Yang–Mills potential as the coordinates of a point in $\mathcal{A}^3$, namely the equivalent of ξ^μ in the notation of the preceding section, with a (the internal symmetry index), $i = 1, 2, 3$ (the spatial index), and x (the point in space) together playing the role of the index μ in ξ^μ.

We are interested in the behaviour of the system under local Yang–Mills gauge transformations. These change the potential $A_i^a(x)$, or in other words, displace points in the base space $\mathcal{A}^3$. We may thus consider the algebra of infinitesimal local (space-dependent) Yang–Mills gauge transformations $\mathcal{G}$ as our displacement algebra acting on points in the base space $\mathcal{A}^3$. Elements of

$\mathcal{G}$ are functions over space taking values in the gauge (structure) group, while elements of $\mathcal{A}^3$ are vector functions over space taking values in the gauge algebra. The displacement algebra and base space we are considering, therefore, are quite different entities. Further, elements of $\mathcal{G}$ can relate only gauge potentials $A_i^a(x)$, or points on $\mathcal{A}^3$, which are gauge transforms of one another. Repeated applications with elements of $\mathcal{G}$ will therefore generate only *gauge orbits*, not an arbitrary path, in $\mathcal{A}^3$. The present gauge structure has thus the features noted above at the beginning of the section.

Elements of $\mathcal{G}$ are matrix-valued functions of x acting on vectors in Γ, and $\mathcal{G}$ has as generators the hermitian matrices $T^a(x)$ satisfying the commutation relations:

$$[T^a(x), T^b(x')] = f^{abc}\delta(x - x')T^c(x'). \qquad (6.3.1)$$

These generators take the place of d_l in (6.2.1) with the labels a and x together playing the role of the index l in d_l.

An infinitesimal gauge transformation on the Yang–Mills field may be represented as $(1 + \epsilon^a(x)T^a(x))$ whose action on $A_i^a(x)$, according to standard rules, is

$$\delta A_i^a(x) = \partial_i \epsilon^a(x) - ig f^{abc} A_i^b(x)\epsilon^c(x), \qquad (6.3.2)$$

or equivalently,

$$\delta A_i^a(x) = \{\nabla_i^{ac}\delta(x - x')\}\epsilon^c(x'), \qquad (6.3.3)$$

where $(\nabla_i)^{ac}$ is the usual covariant derivative:

$$(\nabla_i)^{ac} = \delta^{ac}\frac{\partial}{\partial x^i} - ig f^{abc} A_i^b(x), \qquad (6.3.4)$$

and summation over repeated indices, including x', is understood. This is now to be interpreted as the change in the coordinates $A_i^a(x)$ of a point in the base space $\mathcal{A}^3$ under the action of an element of the displacement algebra, and is to be compared with (6.2.16). The quantity within braces in (6.3.3) is playing the role of the local frames e_l^μ in (6.2.16), with a, i, x together as the index μ, and c, x' together as the index l.

Suppose now we take a wave function Ψ for the whole system of gauge fields plus fermions, which according to the observations above, is just a function of the 'coordinates' $A_i^a(x)$ in our base space. Then following the prescription in (6.3.3), we deduce that the action of the displacement $(1 + \epsilon^a(x)T^a(x))$ on Ψ will result in:

$$\delta\Psi = -\epsilon^a(x)\nabla_i^{ac}\frac{\delta}{\delta A_i^c(x)}\Psi \qquad (6.3.5)$$

where the operator $-\nabla_i^{ac}\delta/\delta A_i^c(x)$ plays the role of d_l in the last section with a and x together as its index.

Consider next the operator:

$$G^a(x) = -\nabla_i^{ac}\frac{\delta}{\delta A_i^c(x)} - igJ^a(x, A),\qquad (6.3.6)$$

introduced by Fadeev (1984), where $J^a(x, A)$ is the left fermion current which specifies an infinitesimal rotation in the Hilbert space Γ of the fermion for each point $A_i^c(x)$ in the base space $\mathcal{A}^3$ and for each direction of displacement $T^a(x)$ therein. We can thus consider $J^a(x, A)$ as a 'connection' or 'potential' analogous to $A^l(\xi)$ of the preceding section, with $A_i^c(x)$ playing the role of the coordinate ξ and a and x together that of the index l. Further, we can interpret $G^a(x)$ in (6.3.6) in our gauge structure as the covariant derivative for the potential $J^a(x, A)$ analogous to D_l of (6.2.5). Following (6.2.11) then, we can write the curvature for the potential $J^a(x, A)$ as:

$$igF^{ab}(x, y) = [G^a(x), G^b(y)] - f^{abc}G^c(x)\delta(x - y),\qquad (6.3.7)$$

where because of (6.3.1), we have substituted $f^{abc}\delta(x - y)$ for the structure constants C_{lk}^m in (6.2.1). But according to Fadeev (1984), the right-hand side of (6.3.7) is the Schwinger term due to the anomaly. We arrive thus at the conclusion that the anomalous term appears as the curvature in the gauge structure under consideration.

Notice that in this example, as in the case of the Berry phase example quoted in the preceding section, the potential is prescribed by the physics. This is different in spirit from the standard formulation of gauge theories where the potential is to be determined by extremizing some functional of the curvature. To effect a complete parallel here, one should consider a situation where the potential, namely the left fermion current $J^a(x, A)$, is to be determined by extremizing in some way a functional of the curvature or anomaly. Although such a viewpoint could in principle be interesting, it has not, as far as we know, been examined in the literature.

6.4   Towards a 'Pointless' Theory

As originally conceived and presented in Chapter 1, gauge theories contain a wave function or field ψ which assigns a phase to each point and a gauge potential A which provides parallel transport of this phase from point to neighbouring point in space-time. However, because of the coincidence there that space-time $\mathbb{R}^4$ is identical to its group of translations, both ψ and A admit an alternative interpretation. Thus, ψ can also be considered as a prescription of how its value changes under translations, and A as a parallel transport of phase under the action of the same. In the generalized framework of the preceding section, we have already advanced from that position; we kept for the gauge potential A only the concept of parallel transport under the action of the displacement algebra, although the wave function there was still capable of either interpretation. We now wish to explore the possibility of discarding also for ψ this dual role, and keep for it only the concept of being a prescription for how its value changes under the action of the displacement algebra. The question is whether a meaningful gauge theory can still be constructed on the broader basis.

If one can do so, then one is led to a physical conception of field theory which is quite revolutionary. According to (6.2.2), the value of ψ at the point ξ' obtained by the action of an infinitesimal displacement on the point ξ is represented by an operator $(1 + \epsilon^l d_l)$ operating on ψ whose value is then taken at the point ξ. The operators d_l were derived originally from the concept of ψ being a function of points in some space, but once the operators d_l are given, one knows already how ψ changes under the action of the displacement algebra without referring again in principle to the points in space from which the operators were originally derived. The same applies also to the gauge potential A_l which are also operators acting on ψ, and which, once specified as such, are sufficient to define parallel phase transport under the action of the displacement algebra, again without referring to the points in the base space. In other words, it seems that the concepts which we want to keep for both ψ and A are already encoded in the operators d_l and A_l. That being the case, can one not perhaps keep only these operators d_l and A_l to construct a gauge theory, and forget about points in the base space altogether? If one could, then one would end up with a 'local' gauge theory without the concept of 'points' to mark locations.

Of course, whether such a tightening of concepts will make physical sense or not will depend on the actual system being treated. For instance, in everyday applications of electrodynamics, we do have, we believe, a conception of space and of time and of the value of a field at points of that space-time, in which case we would not wish to give up that notion. However, the occasion may arise where the physical conditions do not demand such a conception; then such a generalization can be meaningful.

Suppose then that we start with a Hilbert space $\mathcal{H}$ to which the 'wave function' or 'field' ψ belongs. Next, we stipulate that there is an operator algebra $\mathfrak{B}$ acting on $\mathcal{H}$ with generators d_l, which defines what is meant by displacements on ψ. Finally, to specify parallel transport, we introduce a gauge potential A which is to be element of another algebra, say $\mathfrak{A}$, acting on $\mathcal{H}$. With only these as ingredients, can one construct a meaningful gauge theory in analogy with what has gone before? To answer this question, let us retrace the arguments in the first half of Section 6.2 under the new generalized circumstances.

By definition, under an infinitesimal displacement $(1 + \epsilon^l d_l)$, ψ changes to another element in $\mathcal{H}$:

$$\psi \longrightarrow \psi + \epsilon^l d_l \psi, \qquad (6.4.1)$$

while parallel transport under the same displacement leads to the element:

$$\psi \longrightarrow \psi + ig\epsilon^l A_l \psi. \qquad (6.4.2)$$

But how does the gauge potential A_l transform under a gauge transformation? Since we have now no points, we have first to make clear what is meant by a local gauge transformation.

For ordinary gauge theories, as well as for the generalized theories considered in the two preceding sections, a local gauge transformation is parametrized by a matrix-valued function $\Lambda(\xi)$ of the point ξ in the base space taking values in the gauge algebra representing the internal symmetry. In other words, it is an element of the algebra of the matrix-valued functions to which the gauge potential A also belongs, or in our present language, it is an element of $\mathfrak{A}$. The fact that Λ depends on ξ implies that it does not in general commute with the derivative, or in the present language, with elements of the displacement algebra $\mathfrak{B}$. Hence, by abstracting from these considerations, we stipulate that by a 'local' gauge transformation in the new generalized pointless context, we mean a transformation:

$$\psi \longrightarrow \psi' = (1 + ig\Lambda)\psi, \qquad (6.4.3)$$

where ψ belongs to $\mathcal{H}$ and Λ belongs to $\mathfrak{A}$, and Λ does not in general commute with elements of $\mathfrak{B}$.

If we make a displacement now after having performed the gauge transformation (6.4.3), we obtain the wave function:

$$(1 + \epsilon^l d_l)(1 + ig\Lambda)\,\psi = (1 + \epsilon^l d_l)(1 + ig\Lambda)(1 - \epsilon^l d_l)\,\psi^\sim, \qquad (6.4.4)$$

where d_l are the generators of $\mathfrak{B}$ and $\psi^\sim$ is the result (6.4.1) of the displacement before the gauge transformation.

Consider next parallel transport. Before the gauge transformation (6.4.3), we defined the right-hand side of (6.4.2) as being the parallel transport of the wave function ψ under the displacement represented by the action of the operator $(1 + \epsilon^l d_l)$. Because of the gauge transformation, however, the displaced wave function has been redefined as in (6.4.4), so that we should now regard:

$$(1 + \epsilon^l d_l)\,(1 + ig\Lambda)\,(1 - \epsilon^l d_l)\,(1 + ig\epsilon^l A_l)\,\psi \qquad (6.4.5)$$

as the parallel transport of ψ' under the displacement obtained by operating with $(1 + \epsilon^l d_l)$. This same quantity, however, can also be expressed in terms of the gauge transformed potential A'_l by definition as:

$$(1 + ig\epsilon^l A'_l)(1 + ig\Lambda)\psi, \qquad (6.4.6)$$

so that one obtains again the expression (6.2.3) for the transformation law of the gauge potential.

From then on, it is clear that by following step by step the arguments presented in Section 6.2, one can reproduce all the formulae from (6.2.5) to (6.2.15), provided one makes the appropriate changes in the interpretation. For example, using the transformation law (6.2.3), it is easy to show that the expression (6.2.5) for the covariant derivative is still covariant. However, one can of course no longer interpret the covariant derivative operating on ψ as giving the difference between the parallelly transported value and the local value of ψ at the displaced point as one used to do; it has now to be interpreted just as the difference between the value obtained by parallel transport and the displaced value. Further, the field tensor F_{lk} defined as in (6.2.10) is still a covariant curvature of sorts, but it can no longer be interpreted as the change in phase obtained by parallel transport around an infinitesimal closed circuit in the base space. The result, as anticipated above, is a gauge structure in which there is no concept of locality associated with points in a space.

The theory so obtained is constructed by discarding some of the original concepts and requirements, but is the resulting structure necessarily more general that what was considered before? If the algebra $\mathfrak{A}$ is commutative, then the answer is no. As pointed out at the end of Section 3.3, there is a theorem due to Gel'fand which says, roughly, that any commutative algebra can be considered as an algebra of functions over a space, whose derivations are vector fields in that space. Derivations of an algebra, we recall, are linear operations on its elements which obey the Leibnitz rule. Thus Δ is a derivation of $\mathfrak{A}$ if for $a, b \in \mathfrak{A}$,

$$\Delta(ab) = (\Delta a)b + a(\Delta b). \qquad (6.4.7)$$

Now in our considerations above, the generators d_l of $\mathfrak{B}$ act, according to (6.4.4), on the element Λ of $\mathfrak{A}$ by commutation, which satisfies of course the Leibnitz rule (6.4.7), and are therefore derivations of the algebra. Hence, by the theorem cited above, they can be regarded as 'vector fields' or displacements of points in the space over which $\mathfrak{A}$ is an algebra of functions. We are thus back to the situation we were in before in the preceding section. Even if the algebra $\mathfrak{A}$ is noncommutative, of course, the generalized framework under discussion may still reduce to cases previously considered. Ordinary Yang–Mills theory itself is already an example, where the algebra $\mathfrak{A}$ is the algebra of matrix-valued functions over $\mathbb{R}^4$, which is noncommutative.

What physical conditions then distinguish a nontrivial generalization in our present context from the earlier stages? As stipulated above, the displacement algebra $\mathfrak{B}$ acts on elements of $\mathfrak{A}$ by commutation. If $\mathfrak{A}$ is noncommutative, then some elements of $\mathfrak{B}$ can actually be elements of $\mathfrak{A}$ itself, in which case, they are termed *inner derivations* of $\mathfrak{A}$. If that happens, then the theory will be intrinsically different from those we have considered before.

To get a qualitative picture of the conceptual differences involved, let us first examine what happens if $\mathfrak{B}$ contains no element which are inner derivations of $\mathfrak{A}$. Elements of both $\mathfrak{A}$ and $\mathfrak{B}$ being by definition operators acting on the Hilbert space $\mathcal{H}$ of wave functions ψ, we may imagine them to be represented by matrices with rows and columns labelled by various indices enumerating a set of basis vectors in the Hilbert space $\mathcal{H}$. The assertion that no element of $\mathfrak{B}$ is contained in $\mathfrak{A}$ may be interpreted as meaning that the matrices representing elements of $\mathfrak{A}$ are all diagonal with respect to the indices, say, ξ, on which $\mathfrak{B}$ operates. One may then regard the elements of $\mathfrak{A}$ as matrix-valued functions over the space of ξ, which we take as the 'base space' on which the displacement algebra $\mathfrak{B}$ acts. Thus again, we are back to the situation discussed in the preceding section. On the other hand, in the case when $\mathfrak{B}$ contains inner derivations of $\mathfrak{A}$, then they operate also on the indices on which the elements of $\mathfrak{B}$ operate, and hence one can no longer distinguish a base space on which the displacement algebra acts exclusively. The theory will then be 'pointless' in the sense it was meant above.

Note that for the theory to be 'pointless', the requirement is for $\mathfrak{A}$ to be noncommutative but not necessarily $\mathfrak{B}$ to be so. We have developed our formalism since Section 6.2 for a general displacement algebra to stress the fact that it can be noncommutative. However, as far as 'pointlessness' is concerned, so long as $\mathfrak{A}$ is noncommutative so that $\mathfrak{B}$ can contain inner derivations of $\mathfrak{A}$, then it may already not be possible for $\mathfrak{A}$ to be regarded as an algebra of functions over some space, nor $\mathfrak{B}$ as displacements of points in that space, independently of whether $\mathfrak{B}$ itself is commutative or not.

Once we start to deal with pointless gauge theories then even some of our very basic physical concepts which have grown familiar with usage will have to be revised. For example, in all previous cases considered, the wave function ψ depends on two distinct sets of variables — the position of point ξ in the base space, and the internal symmetry index, say i. A gauge transformation acts only on the internal symmetry index i, while displacements of points in the base space acts only on the variables ξ. Thus we have grown used to the conception that the 'internal symmetry space' is disjoint from the base space. That is the reason why, for example, the gauge parameter Λ and the potential A are elements of an algebra of functions and are multiplied and added together like matrices of the internal indices for each value of their argument ξ. To obtain the internal symmetry algebra, we can just take this algebra of functions and factor out the base space from it. However, in the pointless generalizations we are considering, there is first of all no base space as was originally conceived. Consequently, neither can one define an internal symmetry algebra by factoring out the base space from the algebra $\mathfrak{A}$ as we did before. Indeed, there is no longer any 'internal symmetry' as it was previously conceived. The algebra $\mathfrak{A}$ which plays partly that role is now acted on by elements of the algebra $\mathfrak{B}$ which plays the role of displacements. Hence, if we wish to keep the conception of $\mathfrak{B}$ being somehow still associated with displacements in some space, then it means that the symmetry represented by $\mathfrak{A}$ has to be partly associated also with that space. It thus seems that one is dealing here with a gauge theory of an extended object in which the gauge degrees of freedom are at least partly inherent in the object's own spatial extension.

In view of this, it is perhaps not too surprising to find an example of such a gauge structure in the theory of strings. Indeed, it can be shown that string theory which is usually formulated otherwise can also be formulated as a pointless gauge theory of the type described above. The formulation, by virtue of the complexity of the system itself, is a little complicated, and not entirely clear as yet in physical conception. We shall give here only a very brief outline for illustration and refer the interested reader to the literature for further details.

To specify a gauge structure in the pointless framework, it is sufficient to define a Hilbert space $\mathcal{H}$ of wave functions, an operator algebra $\mathfrak{A}$ acting on $\mathcal{H}$ to represent the gauge degrees of freedom, and another operator algebra $\mathfrak{B}$, also acting on $\mathcal{H}$ to represent displacements. It turns out that one will obtain the gauge structure inherent in the standard theory of interacting open bosonic strings if one chooses to specify the three ingredients above respectively as follows. One starts with a function space ΠX, say, of functions $X^\mu(\sigma)$ mapping the interval $[0, \pi/2]$ to d-dimensional space-time as denoted by the coordinates X^μ, $\mu = 0, 1, ..., d-1$, and satisfying the boundary conditions $\partial X^\mu/\partial\sigma = 0$ at $\sigma = 0$, and $X^\mu = x^\mu$ (fixed) at $\sigma = \pi/2$. One then defines the Hilbert space $\mathcal{H}$ as the space of functionals $\Psi[X]$ over this function space ΠX. Next, one takes the operator algebra $\mathfrak{A}$ to be the algebra of (infinite-dimensional) Hermitian matrices $A[X_1; X_2]$ satisfying

$$A^\dagger[X_1; X_2] = A^*[X_2; X_1] = A[X_1, X_2], \tag{6.4.8}$$

where $*$ means complex conjugation and X_1 and X_2 are both elements of the function space ΠX each playing here the role of a continuum of continuous matrix indices. For the displacement algebra $\mathfrak{B}$, one takes then the algebra generated by the operators:

$$L_{\pm\sigma} = \frac{1}{2}\left\{-i\pi\frac{\delta}{\delta X^\mu(\sigma)} \pm X'^\mu(\sigma)\right\}^2, \quad 0 \le \sigma \le \frac{\pi}{2}, \tag{6.4.9}$$

where $'$ denotes differentiation with respect to σ.

That the above assignments for $\mathcal{H}$, $\mathfrak{A}$ and $\mathfrak{B}$ do indeed correspond to the gauge structure in standard string theory can be seen as follows. The generators L_σ of the displacement algebra $\mathfrak{B}$ are to be identified with the operators d_l in (6.2.1) above, only with a continuous index σ in place of the discrete index l. To each displacement operator L_σ then, we should associate an operator A_σ in $\mathfrak{A}$ as the 'gauge potential' to specify parallel 'phase' transport. As noted in (6.2.13) above, the gauge potential can also be written as a differential 1-form:

$$A = \int_{-\pi/2}^{\pi/2} d\sigma\, A_\sigma \eta^\sigma, \tag{6.4.10}$$

by introducing the anti-commuting 'differential 1-forms' η^σ, and replacing the summation over l in (6.2.13) by an integral over the continuous index σ in (6.4.10). That being the case, then according to (6.2.15), the exterior derivative can be written as:

$$Q = \int_{-\pi/2}^{\pi/2} d\sigma \left\{ [L_\sigma,\ ]\eta^\sigma + 4i\pi\eta^\sigma\eta'^\sigma \frac{\delta}{\delta\eta^\sigma} \right\}. \tag{6.4.11}$$

where we have substituted for the structure constants C_{lk}^m in (6.2.15) and (6.2.1) the corresponding quantities for our present displacement algebra $\mathfrak{B}$:

$$C_{\sigma_1\sigma_2}^{\sigma_3} = 4i\pi\delta'(\sigma_1 - \sigma_2)[\delta(\sigma_3 - \sigma_1) + \delta(\sigma_3 - \sigma_2)], \tag{6.4.12}$$

which can be directly computed from (6.4.9) by commuting with the operators L_σ. With the exterior derivative (6.4.11), one can then construct as usual, from the potential 1-form, gauge covariant quantities, such as the curvature form (6.2.14), and hence also gauge invariants by taking traces. One example of a possible invariant in the present case is the trace of the Chern-Simons 3-form:

$$\mathcal{A} = Tr(AQA + \frac{2}{3}AAA), \tag{6.4.13}$$

which will turn out to be of particular interest.

Introduce next the notation:

$$\mathbf{X}(\sigma) = X_1(\sigma),\ 0 \le \sigma \le \pi/2; \quad \mathbf{X}(\sigma) = X_2(\pi - \sigma),\ \pi/2 \le \sigma \le \pi, \tag{6.4.14}$$

so that the potential 1-form in (6.4.10) may now be considered as a functional $\mathbf{A}[\mathbf{X}]$ of the 'full-string' $\mathbf{X}(\sigma)$, $\sigma = 0 \to \pi$, instead of as a matrix $A[X_1; X_2]$ of the 'half-strings' $X_1(\sigma)$ and $X_2(\sigma)$, $\sigma = 0 \to \pi/2$. It is in fact this $\mathbf{A}[\mathbf{X}]$ which is to be identified with the string functional in the conventional formulation of string theory. By making the appropriate changes in notation it can then be shown that the operator Q in (6.4.11) operating on the matrix $A[X_1; X_2]$ is in fact the same as the operator:

$$\mathbf{Q} = \int_{-\pi}^{\pi} d\sigma \left\{ \boldsymbol{\eta}^\sigma \mathbf{L}_\sigma + 4i\pi \boldsymbol{\eta}^\sigma \boldsymbol{\eta}'^\sigma \frac{\delta}{\delta \boldsymbol{\eta}^\sigma} \right\} \qquad (6.4.15)$$

operating on the full-string functional $\mathbf{A}[\mathbf{X}]$. Readers familiar with string theory will immediately recognize $\mathbf{Q}$ in (6.4.15) as the BRST charge operator of the conventional formalism with $\mathbf{L}_\sigma$ as the Virasoro operators and the anti-commuting differentials $\boldsymbol{\eta}^\sigma$ as the BRST ghosts. Indeed, when transcribed to this notation, the gauge invariant (6.4.13) is seen to be exactly Witten's action in the conventional formulation of string field theory.

Proceeding in this way, one sees that one can recover the standard open bosonic string theory, but now as a 'pointless' gauge theory of the type described above in this section. Although in the present outline, we have been extremely cavalier in handling infinite-dimensional quantities, the equivalence between the standard formulation and the one above is known to remain valid in a more careful treatment when, for example, ordering of mode oscillators are properly taken into account, and when quantities labelled by continuum indices are avoided. We shall not, however, go into these detailed questions of rigour, since what interests us here is not string theory for its own sake, but as an example of the pointless generalization of Yang–Mills theory we have been considering.

Indeed, when considered as such, string theory neatly exhibits all the unusual features of pointless gauge theories that we have discussed. First, one sees that string theory is 'pointless' in that one cannot identify a base space over which one can describe the theory as a local gauge theory in the normal sense. True, the wave function $\Psi[X]$ is a functional over the space ΠX of functions $X(\sigma), \sigma = 0 \to \pi/2$, so that one may be tempted to think of ΠX as the base space for the theory. However, one sees that the algebra $\mathfrak{B}$ which plays the role of displacements here has as generators L_σ of (6.4.9), which though consisting of an odd part:

$$L_\sigma - L_{-\sigma} = -2i\pi X'^\mu(\sigma) \frac{\delta}{\delta X^\mu(\sigma)}, \qquad (6.4.16)$$

which is a tangent vector, or a genuine displacement of points in the space

ΠX, it also contains an even part:

$$L_\sigma + L_{-\sigma} = -\pi^2 \frac{\delta}{\delta X'^2(\sigma)} + X'^2(\sigma), \qquad (6.4.17)$$

which is nothing of the sort. This is in contradiction with what we have understood so far as the relation between the base space and the displacement algebra of a local gauge theory. Second, one is also unable to identify for string theory an internal symmetry group in the sense it was usually understood. The algebra $\mathfrak{A}$ representing the gauge degrees of freedom is here the algebra of Hermitian matrices $A[X_1; X_2]$ where X_1 and X_2 are in ΠX. By analogy with usual gauge theories, one may then be tempted to identify the internal symmetry group as the group of unitary matrices $U[X_1; X_2]$ satisfying:

$$U^\dagger U = \int \delta X_2 U^*[X_2; X_1] U[X_2; X_3]\, \delta[X_1 - X_3]. \qquad (6.4.18)$$

In other words, one may be tempted to treat the functions X_1, X_2 etc. as internal symmetry indices just like, say, the three-valued colour indices of QCD. In QCD, however, the elements of the displacement algebra, namely the generators of space-time translations ∂_μ, do not act on the colour indices of the internal symmetry; whereas, here in string theory, the generators L_σ of $\mathfrak{B}$ act on the indices X_i as well. Indeed, as already noted, it is a peculiar characteristic of pointless gauge theories that our old familiar concepts of internal symmetry and space-time translations, which were completely disjoint before, has here become completely intertwined.

Regardless of whether one shares the once-popular enthusiasm for string theory and the ambitious hope that it may be a model 'theory of everything', one sees that it has served to illustrate the subtle and revolutionary changes in physical concepts required in generalizing from the usual local gauge theory to a 'pointless' theory. Although one may not have grasped as yet their full physical significance, nor yet known concrete examples where they are of practical use, the new concepts seem to open new horizons which could be of fundamental importance and wide applicability in the future.

Bibliography

CHAPTER 1

Aharonov, Y. and Bohm, D. (1959).
Significance of electromagnetic potentials in the quantum theory.
Phys. Rev. **115**, 485–491.
Chambers, R.G. (1960).
Shift of an electron interference pattern by enclosed magnetic flux.
Phys. Rev. Lett. **5**, 3–5.
Chan Hong-Mo and Tsou Sheung Tsun (1981).
Monopole charges in unified gauge theories.
Nucl. Phys. **B189**, 364–380.
Dirac, P.A.M. (1931).
Quantised singularities in the electromagnetic field.
Proc. Roy. Soc. London **A133**, 60–72.
Wilson, K. G. (1974).
Confinement of quarks.
Phys. Rev. **D10**, 2445–2459.
Wu, Tai Tsun and Yang, Chen Ning (1975).
Some remarks about unquantized non-Abelian gauge fields.
Phys. Rev. **D12**, 3843–3844.
Wu, Tai Tsun and Yang, Chen Ning (1975).
Concept of nonintegrable phase factors and global formulation of gauge fields.
Phys. Rev. **D12**, 3845–3857.
Yang, Chen Ning (1979).
Charge quantization, compactness of the gauge group, and flux quantization.
Phy. Rev. **D1**, 2360.
Yang, Chen Ning and Mills, R. L. (1954).
Conservation of isotopic spin and isotopic gauge invariance.
Phys. Rev. **96**, 191–195.

Chapter 2

Chan Hong-Mo and Tsou Sheung Tsun (1981).
  Monopole charges in unified gauge theories.
  Nucl. Phys. **B189**, 364–380.
Chan Hong-Mo and Tsou Sheung Tsun (1981).
  On the confinement of monopoles in non-abelian gauge theories.
  Rutherford Laboratory report RL–81–031, unpublished.
Coleman, S. (1974).
  Classical lumps and their quantum descendants.
  Erice School, p. 297.
Dirac, P.A.M. (1931).
  Quantised singularities in the electromagnetic field.
  Proc. Roy. Soc. London **A133**, 60–72.
Lubkin, E. (1963).
  Geometric definition of gauge invariance.
  Ann. Phys. (New York) **23**, 233–283.
Nambu, Y. (1977).
  String-like configurations in the Weinberg–Salam theory.
  Nucl. Phys. **B130**, 506–515.
Nielsen, H. B. and Olesen, P. (1973).
  Vortex lines models for dual strings.
  Nucl. Phys. **B61**, 45–61.
Polyakov, A. M. (1975).
  Isomeric states of quantum fields.
  Sov. Phys. JETP **41**, 988–995.
  Russian original *Zh. Eksp. Teor. Fiz.* **68**, 1975–1990.
't Hooft, G. (1974).
  Magnetic monopoles in unified gauge theories.
  Nucl. Phys. **B79**, 276–284.
't Hooft, G. (1978).
  On the phase transition towards permanent quark confinement.
  Nucl. Phys. **B138**, 1–25.
Wu, Tai Tsun and Yang, Chen Ning (1975).
  Concept of nonintegrable phase factors and global formulation of gauge
  fields.
  Phys. Rev. **D12**, 3845–3857.

CHAPTER 3

Borel, A. (1955).
Topology of Lie groups and characteristic classes.
Bull. Amer. Math. Soc. **61**, 397–432.
Bott, R. and Tu, L. W. (1982).
Differential Forms in Algebraic Topology.
Springer–Verlag, New York.
Chern, S. S. (1968).
Complex Manifolds Without Potential Theory.
Springer–Verlag, New York.
Connes, A. (1985).
Non-commutative differential geometry.
Publ. Math. IHES **62**, 41–144.
Fulton, W. and Harris, J. (1991).
Representation Theory.
Springer–Verlag, New York.
Göckeler, M. and Schücker, T. (1978).
Differential geometry, gauge theories, and gravity.
Cambridge University Press, Cambridge.
Helgason, S. (1978).
Differential Geometry, Lie Groups and Symmetric Spaces.
Academic Press, San Diego.
Isham, C. J. (1989).
Modern Differential Geometry for Physicists.
World Scientific Publishing Co., Singapore.
Kobayashi, S. and Nomizu, K. (1963).
Foundations of Differential Geometry, Vol. 1.
Interscience Publishers, New York.
Pressley, A. and Segal, G. (1986).
Loop Groups.
Clarendon Press, Oxford.
Price, J. (1977).
Lie Groups and Compact Groups.
Cambridge University Press, Cambridge.
Samelson, H. (1969).
Notes on Lie Algebras.
Van Nostrand Reinhold Co., New York.
Steenrod, N. (1951).
The Topology of Fibre Bundles.
Princeton University Press, Princeton.

CHAPTER 4

Barrett, J. W. (1990).
  Holonomy and path structures in general relativity and Yang–Mills theory.
  Int. J. Theor. Phys. **30**, 1171–1215.
Chan Hong-Mo, Scharbach, P. and Tsou Sheung Tsun (1986).
  On loop space formulation of gauge theories.
  Ann. Phys. (New York) **166**, 396–421.
Corrigan, E. and Hasslacher, B. (1979).
  A functional equation for exponential loop integration in gauge theories.
  Phys. Lett. **B81**, 181–184.
Mandelstam, S. (1979).
  Charge–monopole duality and the phase of non-abelian gauge theories.
  Phys. Rev. **D19**, 2391–2409.
Polyakov, A, M. (1980).
  Gauge fields as rings of glue.
  Nucl. Phys. **B164**, 171–188.
Wu, Tai Tsun and Yang, Chen Ning (1975).
  Concept of nonintegrable phase factors and global formulation of gauge
  fields.
  Phys. Rev. **D12**, 3845–3857.

CHAPTER 5

Chan Hong-Mo, Scharbach, P. and Tsou Sheung Tsun (1986).
  Action principle and equations of motion for nonabelian monopoles.
  Ann. Phys. (New York) **167**, 454–472.
Chan Hong-Mo and Tsou Sheung Tsun (1993).
  Quantum mechanics of Dirac–like topological charges in Yang–Mills fields.
  In preparation.
Polyakov, A, M. (1980).
  Gauge fields as rings of glue.
  Nucl. Phys. **B164**, 171–188.
Wong, S. K. (1970).
  Field and particle equations for the classical Yang–Mills field and particles
  with isospin.
  Nuovo Cim. **65A**, 689–694.
Wu, Tai Tsun and Yang, Chen Ning (1976).
  Dirac monopoles without strings: classical Lagrangian theory.
  Phys. Rev. **D14**, 437–445.

CHAPTER 6

Berry, M. V. (1984).
Quantal phase factors accompanying adiabatic changes.
Proc. Roy. Soc. London **A392**, 45–57.
Birse, M. C. and McGovern, J. A. (1988).
Cranking, adiabatic phases and monopoles.
J. Phys. **A21**, 2253–2260.
Cartan, E. and Schouten, J. A. (1926).
On the geoometry of the group-manifold of simple and semi-simple groups.
Proc. Skad. Wetensch. Amsterdam **29**, 803–815.
Chan Hong-Mo and Tsou Sheung Tsun (1989).
Yang–Mills formulation of interacting strings.
Phys. Rev. **D39**, 555–564.
Chan Hong-Mo and Tsou Sheung Tsun (1990).
Towards a 'pointless' generalization of Yang–Mills theory.
Ann. Phys. (New York) **198**, 180–200.
Connes, A. (1985).
Non-commutative differential geometry.
Publ. Math. IHES **62**, 41–144.
Fadeev, L. D. (1984).
Operator anomaly for the Gauss law.
Phys. Lett. **B145**, 81–84.
Gervais, J.-L. (1986).
Gauge invariance over a group as the first principle of interacting string dynamics.
Nucl. Phys. **B276**, 349–365.
Nelson, P. and Alvarez-Gaumé, L. (1985).
Hamiltonian interpretation oof anomalies.
Comm. Math. Phys. **99**, 103–114.
von Neumann, J.
Collected Works, Vol. IV, p. 127.
Pergamon Press, Oxford 1962.
Witten, E, (1986).
Noncommutative geometry and string field theory.
Nucl. Phys. **B268**, 253–294.

Index